Guide for the
Perplexed Organic
Experimentalist

H. J. E. LOEWENTHAL

**Chemistry Department,
Technion—Israel Institute
of Technology**

HEYDEN

London . Philadelphia . Rheine

Heyden & Son Ltd., Spectrum House, Hillview Gardens, London
NW4 2JQ, UK
Heyden & Son Inc., 247 South 41st Street, Philadelphia, PA 19104,
USA
Heyden & Son GmbH, Münsterstrasse 22, 4440 Rheine/Westf.,
Germany

ISBN 0 85501 169 6

Type set by John Wright & Sons Ltd, Bristol and printed in Great
Britain by Henry Ling Ltd, The Dorset Press, Dorchester

Contents

Preface

The perplexed organic experimentalist is in my experience the beginning research student and (frequently) the post-doctoral research worker. He is the one who early on discovers that he has to stand on his own two feet, in the task of searching for information on his subject and in all the practical aspects of his work— and that means not only how to run a reaction but also how to choose and acquire the tools and materials of his trade.

All too often it is not by choice that he finds himself in this situation. His supervisor (or, more politely, his 'Senior Collaborator') was himself once a graduate student and post-doctoral researcher. But in the majority of cases he has long since abandoned the laboratory bench and is now busy with administration, with writing and refereeing research grant applications and scientific papers, and with teaching and thinking. In the process he will have forgotten most of the practical knowledge which he had to acquire painfully in his own day and will have become unaware of later developments.

The average graduate student is ill-prepared for searching the literature. Most practical textbooks will have done little to train him to think for himself. Few prepare him for continuing preparative work on a small scale, and fewer still for working with sensitive reagents and under dry and anaerobic conditions. None, so far as I know, do anything to assist him (or his supervisor) to grapple with indifferent suppliers, manufacturers and administrators.

This small book attempts to fill some of the gaps, on the basis that organic chemistry is an experimental science first and foremost, and that in the final analysis—with all of modern instrumentation and computerization—it is the organic chemist's brain and own two hands that are and will remain indispensable. Hence there is little on instrumental spectroscopic and analytical techniques and other special fields, which are quite adequately dealt with elsewhere. The important matter of safety is, with regret, touched upon only in brief; to do proper justice to it would be beyond the scope of this book. Workers in certain areas such as carbohydrates or peptides may find comparatively little of special relevance to them. Others will no doubt take exception to a number of statements in this book. I shall be happy if their misgivings will induce them all to write a similar book of their own.

I am indebted to many friends and colleagues for constructive advice and criticism, and above all to Professor R. A. Raphael, F.R.S. (University of Cambridge), for his encouragement and great help, and to Professor O. Jeger (Swiss Federal Institute of Technology, Zurich) whose hospitality enabled me to get started on this work.

Haifa,
January 1978   H. J. E. LOEWENTHAL

1
On Searching the Literature

This chapter comes first because that is where it ought to be. No serious programme of research, chemical or otherwise, is ever embarked upon before a thorough investigation of the current state of knowledge of the subject has been undertaken.

It is instructive to examine the way in which the use of the primary and secondary literature is treated in the teaching of organic chemistry nowadays. Inspection of 30 textbooks in English, published or revised since 1960, shows that 25 of them make practically no mention of the subject at all and of these 15 do not even give a bibliography or reading list. All the remaining five relegate it to the end of the book, which will hardly impress its importance upon the student. Only three of the latter make a serious effort to guide the student in the use of that literature. These findings, incidentally, are unrelated to whether the book is called simply 'Organic Chemistry' or whether the title is qualified by terms such as 'Basic', 'Introduction to . . .',

'Modern', 'Comprehensive' or 'Advanced'. Surely this state of affairs ought to raise some searching questions on the education of the organic chemist in the age of the information explosion.

There are books which deal with the subject specifically; among these the ones by Crane et al.,[1] by Dyson,[2] by Bottle,[3] by Burman,[4] and the A.C.S. Monograph on the subject[5] can be mentioned; and anyone wanting to study the subject in depth should use these. However, they are designed to cover the literature of chemistry as a whole, and none are up to the latest developments.

The object of this chapter is to provide some hints and guidance to the beginning organic research chemist, and in particular to point out some of the snags and pitfalls he is bound to encounter.

SEARCHING FOR A COMPOUND, ITS SYNTHESIS AND PROPERTIES

The most frequent object of a literature search consists of finding out whether a certain compound is known, how it was prepared and what its properties are (or appear to be). In this, and indeed in any other search, one tries to find everything known in the past up to the point where one's own work should start. In fact, in most cases the search will determine just what that starting point should be.

It should follow that any such search should be made in reverse chronological order, i.e. starting with the latest Index of *Chemical Abstracts*, the most recently issued series of *Beilstein's Handbuch* or the latest review, and then to work backwards in time on the reasonable assumption that whosoever worked last on the point in question will have gathered most if not all the previous information for you. As obvious as this may seem it is mentioned all the same; no one seems to have stressed it particularly before and in fact much of the customary advice (as much as there is of it) implies doing the whole process in the opposite direction. This does not mean that having arrived at the first past author who appears to have reviewed prior endeavours reasonably well you should consider your work done. As in driving a car, always consider the other fellow worse than you are and then set out to prove it. In most cases, if you dig deep enough you will surely discover some piece of prior and important information which he has missed and which will put you at an immediate advantage.

The beginner is tempted to waste his time placing too much reliance on various handbooks, compendia, 'dictionaries' and 'encyclopedias' found in most libraries and laboratories. For any serious search these are incomplete and liable to be inaccurate, and where literature references are given it is not always clear just to what particular point they do refer.

Chemical Abstracts

This is the most important source of information on chemical compounds, at any rate from the present back to the year 1949 when in the majority of cases *Beilstein's Handbuch* should take over (see below). For this reason that part of any library which houses this work is the most frequented one and that area should be reserved for *Chemical Abstracts* readers only. It should also go without saying that no volume of this work should ever leave that area and that its binding should receive special care.

The most important parts are the *Indexes*: the *Collective* ones in so far as they have appeared and the *Semi-annual* ones if not. Since 1971 the *Subject Indexes* have been sub-divided into a *General Subject* and a *Chemical Substance Index*. Of comparable importance are the *Formula Indexes*; the *Author Index* is of course of secondary importance in this connection.

The main problems encountered by the beginning searcher are those of Nomenclature, of Evaluating the Entry and of Evaluating the Abstract.

Nomenclature

In this *Chemical Abstracts* have gone their own way over the years, undaunted by IUPAC and other rules. The ever-increasing use of computer search and printing does of course provide a generally convincing rationale for this.

A few examples will serve to show what has happened in the course of time:

(a) Up to 1946: α-Toluic acid
     1947–71   : Acetic acid, phenyl
     Now   : Benzeneacetic acid
(b) Up to 1971: Acetic acid, chloroformyl, ethyl ester
     Now   : Propanoic acid, 3-chloro-3-oxo, ethyl ester
(c) Up to 1971: o-Anisidine
     Now   : Benzeneamine, 2-methoxy
(d) Up to 1971: Norephedrine
     Now   : Benzenemethanol, α-(1-aminoethyl)
(e) Up to 1946: Alanine, N-acetyl-β-phenyl
     1947–71   : Alanine, N-acetyl-3-phenyl
     Now   : Phenylalanine, N-acetyl

Compare with the corresponding N-phthaloyl derivative:

(f) Up to 1971: 2-Isoindolineacetic acid, α-benzyl-1,3-dioxo
     Now   : 2H-Isoindole-2-acetic acid, 1,3-dihydro-1,3-
                    dioxo-α-(phenylmethyl)

The tendency of the changes introduced recently is ruthlessly to cut out names considered 'trivial' and thus eliminate ambiguity as far as possible. This is indeed true in a case such as example (b). When searched for in previous years in the Formula Index under $C_5H_7ClO_3$ one would have come across something like 'malonaldehydic acid, chloro, ethyl ester', which at least at first glance would have looked like being the right thing. The new name, unreasonable as it seems and indeed is from a functional point of view, leaves no room for doubt. As against that the new 'Phenylalanine' [example (e)], while perhaps 'generally accepted' and hence 'justifiably trivial' is certainly not un-ambiguous.

One could go on and on with further examples of often bizarre changes in nomenclature, but the above should suffice to bring home the importance of being on one's guard all the time. Fortunately the latest changes have been accompanied by the issue of a handy and separate Index Guide which copiously lists and cross-references all new names and old. *This must be used when in any doubt whatever*—all too often a literature search turns out to be woefully incomplete simply because changes in nomenclature have not been taken into account.

The beginner, not wishing to be bothered with all these complications, tends to head straight for the *Formula Index* which he thinks will circumvent them. He should be disillusioned as early as possible. For instance, let him search there for a specific compound of empirical formula, say, $C_{10}H_{14}O$, and then let us see how he will cope with wading through, and choosing the right one from, more than 500 alternatives. Or, suppose he has to look for a hydroxy-dicarboxylic acid $C_{32}H_{50}O_5$ and compounds related to it. He will probably and eventually find the one he is looking for, but what about its mono- and dimethyl or ethyl or benzyl esters, the corresponding O-acetates, -benzoates or -propionates, any one or more of which may be better known, more completely characterized or more easily accessible than the parent acid? In the *Subject* or *Chemical Substance Index* they will all be grouped together, but in the *Formula Index* each one of them has to be looked for separately, and may even not be found at all since it is known that this Index is often incomplete.

Here one ought to mention another aid to compound search which by contrast the beginner seems to avoid as much as he can even though it is quite simple to use: the *Ring Index*. This lists the systematic names of ring systems according to (a) the number of discernible rings, (b) the size of each, and (c) the nature and number of heteroatoms in each ring—arranged according to the alphabetical Hill system. To give examples: (i) the following compound

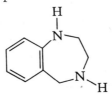

has two rings, one six- and the other seven-membered, the first has only carbons atoms and the second two nitrogen atoms.

It will appear in the ring index under: two rings; 6,7; C_6—C_5N_2, and the corresponding basic system will appear under the name 1,3-Benzodiazepine. (ii) A more complex example:

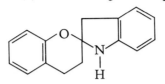

has four discernible rings, will be classed as: 5,6,6,6 and C_4N—C_5O—C_6—C_6, and following up the Ring Index under this will be found the name Spiro [2H-1-benzopyran-2,2′-indoline]. (iii) A steroid having a two-oxygen bridge between positions 5 and 7 will on inspection be seen to incorporate five rings; and since the two-oxygen bridge is common to two of them the compound must be classed as C_5—C_4O_2—C_4O_2—C_6—C_6. A possible and ultimately found to be correct candidate is 'cholestane, 5,7-epidioxy'. The old *Ring Index*, published separately with supplements until the early sixties, had the advantage of showing actual structures in each instance. Since 1962 it has formed part of the *Chemical Abstracts* Indexes (starting from the *Collective* one for 1962–66); and for structures no longer being shown it is necessary to refer back and find the appropriate one in the *Subject* or *Chemical Substance Index*.

Evaluating the entry

This and the following section bring home the fact that hundreds of people are involved in preparing *Chemical Abstracts* and that human nature and subjectivity can play a part.

The entry for a compound occasionally carries no topic, but generally one is given in abbreviated form. The art lies in interpreting its wording alone or in context, and thus deciding which entries are more or less relevant and which abstract is worth or not worth looking at. Of course practice makes perfect, but herewith a few hints to get you started assuming, for example, that you are looking for a compound's synthesis:

1. Under the compound are also listed its functional derivatives (esters, salts, complexes, oximes, hydrazones, etc.). The

relevance of these will depend on how easily these derivatives can be converted into the desired compound or whether they can serve as a reasonable substitute. Further evaluation should proceed as below.

2. A number of entries may refer to the same abstract. This may be significant on its own and making a note of it immediately will save time.

3. No topic: in the end this may be your best bet. Often it means that the paper carries so much information that the abstracter and indexer preferred not giving a topic at all.

4. 'preparation of': in my experience this is prone to refer to a method by which the compound was indeed prepared but which turns out to have little preparative value.

5. 'raman spectrum of', 'nuclear magnetic resonance spectrum of', 'dipole moment of': if this is an unduplicated entry it may refer to the work of someone (spectroscopist or other physical chemist) who obtained the small amount of compound required from a colleague and was possibly too busy arranging his data to make it or even give a literature reference to it.

6. 'bromination of', 'oxidation of', 'reaction of, with phenyl-magnesium bromide': an excellent prospect; clearly work done by an organic chemist who had to make the compound himself in larger quantities.

7. 'carcinogenic activity of': possibly useless.

8. 'in grapefruit seed': probably useless.

9. 'bond length in': utterly useless.

Evaluating the abstract

There was a time when the abstract would refer directly to all new compounds made, their melting or boiling points and possibly other physical properties, and outline details of their preparation. Evaluation was then based in the main on how trustworthy one considered the authors and the journal concerned. This is a subject on which I prefer not to make any comment at all here. But those days have gone and the average abstract, often leaning heavily on the author's own summary, has become not much more than another shunting point *en route* to the original publication. Evaluation is thus reduced to choosing which journal to look at at all, and which to look at first. If it happens to be found in your own library the decision is simple. If not, the question arises whether to take the next train to wherever publications such as *Farmaco, Ed. Sci.*; *Khim. Geterosikli Soedin.* or *Sb. Pr. Pedagog. Fak. Ostrave Rada E.* can be found. This should not even be contemplated unless you know the language; and ordering a Xerox copy having located someone who does is not always a solution. There are few experiences as frustrating as trying to translate an article in Yakugaku Zashhi with the help of a Japanese lecturer of Social Economics.

An important thing to ascertain is whether the article is in the form of a Communication and thus liable to give little experimental detail. If reference is to *Angew. Chem.*, *Chem. Lett.*, *Chimia*, *J. Chem. Soc.*, *Chem. Commun.* or *Tetrahedron Letters* the situation is clear, but not if to *J. Am. Chem. Soc.* or *J. Org. Chem.* The last two are likely to give a modicum of experimental facts, the first four less so. As far as *J. Am. Chem. Soc.* is concerned you will soon find out that certain authors publish *nothing but* Communications or rarely a full paper. Here the prospect of ever finding fuller detail is not bright at all unless the senior author will condescend to send it to you after you have written to him personally.*

The fact that the reference is to a journal which from its title is not strictly in the organic chemistry field need not deter you. Some of the best and most detailed information on synthesis and preparation can turn up in journals such as *Agric. Biol. Chem.*, *Phytochemistry* and *J. Organomet. Chem.* which cover adjacent fields in which the majority of authors started their career as organic chemists, and it is only natural for them to show that they have not lost their touch and expertise.

* And when you write do not forget to check whether there has not been a change in address from that given in the article, lest you commit the unpardonable offence of not knowing that the man has moved to a more prestigious institution.

And now a few comments on cases where the abstract is of a patent. This is of course a special problem when you are nowhere near a patent library, and when the abstract gives no detail worthy of consideration. By and large I would say that it is definitely advisable to order (by air mail) a copy of the patent, assuming you have made quite sure that the information in question has not subsequently been published in the regular form (and this does happen). I do not think that automatically turning up one's nose at a patent reference is justified, and I can think of a number of cases in my own experience where in the preparation of a rather tricky kind of compound following a patent, and not a paper in a reputable journal, led to success. Moreover, there are many fields and whole groups of compounds which because of incidental commercial importance are not covered anywhere *but* in the patent literature. In the end, however, you should keep in mind always that patents are written by people who first and foremost are lawyers and not scientists, and that their prime object is not dissemination of knowledge for its own sake but the staking of a legal claim—and this usually by not giving away more information than is absolutely necessary for that specific purpose.

Beilstein's Handbuch

This is a unique work, and no self-respecting organic chemist can afford to be unfamiliar with its use. It is of course written in German, but once the English-speaking reader has come to terms with this he cannot but admire the perfectly thorough and logical system by which information is organized, and which has not changed from the early beginning to the present day—this is perhaps the most outstanding characteristic of the work. An excellent starting point for getting aquainted with it is the schematic diagram which appears in the textbook by Hendrickson *et al*.[6] and the chapter by Owen and Rickett[7] which appears in Bottle's book; after this the recent guide byWeissbach[8] should be consulted; this incidentally contains a useful glossary of German words and their translation into English and French.

'*Beilstein*' is organized into acyclic (Vols 1–4), isocyclic (carbo-cyclic) (Vols 5–16) and heterocyclic (Vols 17–27) parts. There are some additional but now insignificant volumes. In each Volume the original version (*Hauptwerk*, H) has been complemented by additional series (*Ergaenzungsbände*, EI, EII), and the three series are complete up to 1929. Acyclic and carbocyclic compounds are further covered up to 1949 in a third complementary series (EIII), and a few heterocyclic systems up to 1959 in a combined third and fourth (EIV) one. A few acyclic systems are also covered in a fourth series up to 1959 but you can count yourself 'lucky' for having occasion to use this.

Hence, as already mentioned, in the majority of cases (i.e. all compounds other than heterocyclic) '*Beilstein*' can and should take over from *Chemical Abstracts* for any information prior to 1950, and of course there is no comparison as far as wealth of detail is concerned. The principle of searching in reverse chronological order is also appropriate for '*Beilstein*'. Each series refers back in each case to the appropriate page in previous series (H, EI, EII) if the compound was in any way mentioned there, and also to any previous error. The Indexes (the combined General and Formula ones for (H + EI + EII) and the separate ones for each volume of EIII and EIV) make allowances for many nomenclature nuances, quite unlike *Chemical Abstracts*. An important auxiliary is the *System Number* allotted to each com-pound and its functional and homologous modifications. This continues unchanged through all the series and helps in locating the appropriate book particularly in the later series, though for this it is occasionally necessary to determine it through a similar or homologous compound in the earlier *General Index*.

The great value of '*Beilstein*' lies not only in the mass of information it provides but also in its critical approach to the literature. Not everything reported is taken for granted and later series never hesitate to correct errors in previous ones. Then there is the gold mine of hundreds of compounds of known empirical formula but of unknown constitution, some of which turn out later to be of real importance—and only a conscientious '*Beilstein*' reader can spot them Many of them still await the organic chemist who will solve their structure by modern instrumental methods in a matter of days (and perhaps 'benefit' by a quick publication as a result).

Above all—you *must* know the system. Only then will you be able to decide whether an amide or an ester is to be found under the amine or alcohol or under the acid, or to avoid the frustration from discovering that succinic anhydride is nowhere near succinic acid but has been relegated to the volume marked Heterocycles, 1,O, mono- and polyoxo- compounds. But in the end you should be grateful that there *is* a system.

Some Other Recommended Sources for Compound Search

Elsevier's *Encyclopedia of Organic Chemistry.*[9] This was a valiant attempt to create an English 'Beilstein' which fizzled out. The 18 volumes which did appear cover only certain carbobicyclic and -polycyclic systems, for example indanes, bicyclo- and spiro- alkanes, naphthalene compounds, anthracenes and higher poly- cyclics, steroids. Those dealing with naphthalenes and steroids are especially useful. The coverage is now rather out of date; in most cases only up to 1945 though in some fields there is further partial coverage up to 1961. But for anyone who happens to work in the fields covered the work is a pleasure to use. The printing and structural formulae are admirable, and the system used is crystal-clear the moment one sees the list of contents. Other good points are the indexes, and the way references are listed at the end of each section with their location indicated at the bottom of nearly every page.

Pouvoir Rotatoire Naturel, 1a. Stéroides.[10] This happens to be still the only reasonably complete list of steroids and hence aid for compound search in this area. The latest 1965 edition covers the literature up to 1961. The only specific information given on each compound is its optical rotation, but the references to it enable one to acquire further knowledge. Compounds are listed in order of empirical formula but at the back there is also an alphabetical name index. Similar but more out-of-date volumes have appeared for triterpenes and alkaloids.

W. Karrer's *Konstitution und Vorkommen der Organischen Pflanzenstoffe*.[11] This is a convenient reference work on compounds occurring in plants (excepting alkaloids). Coverage is only up to 1956. Compounds are arranged, logically enough, by functional groups and the literature references given under each compound are accompanied in most cases by a brief explanation. The index is divided into a chemical and a botanical part.

SEARCHING FOR A REACTION, METHOD, CONCEPT, PROCESS

This is much more difficult because the search object is by nature less well defined. If it is a name reaction (Reformatski), a standard method (electrophoresis) or accepted concept (neighbouring group participation) the task is relatively simple. It will become progressively more difficult as its designation becomes more complex and involved (Michael-type intramolecular cyclization, gas chromatography of silylated amino acids, effect of solvents on enolization).

There are two possible approaches to the problem. One is to remember that most such general search objects can be expressed in a number of ways, and that the abstracter's or indexer's ideas on this will frequently be quite different from yours. Much will depend on your ability to put yourself in his place; and it is in any case a salutary and educational experience to sit down first and write up all the possible variations in which your problem can be formulated. You will, however, need a lot more initiative and imagination than merely varying the order of key words in a long sentence—any computer can do that.

The second is to see how much information can be gathered by splitting up the problem into component parts and examining each closely. For example, when studying the possibility of getting compound B from A using a reaction such as R and possibly a reagent such as X, one should (a) look up every known review on R and see if their tabular surveys show anything like

A as starting material and like B as product, (b) look up compound A in, for example, *Beilstein's Handbuch* and see whether it has ever been subjected to a similar reaction, (c) look up B and see whether it has ever been obtained in a similar way, (d) do the same with compounds functionally similar to, but in the past more accessible than, A and B, (e) look up X and similar compounds in books devoted to reagents to see whether their use has ever been called upon in connection with analogues of A and B, and so forth.

Obviously, whichever approach you use first or at all you will be engaged in much toing and froing, trying to assemble, tie up and disentangle items and ideas from many different sources. A searcher of this type can be spotted from afar by the glazed and yet determined look in his eyes and his failure to recognize even close aquaintances. One compensation you are bound to acquire will come from the large amount of incidental, unconnected and yet potentially useful information you will pick up on the way, but in this as in the search itself the benefit you will derive will depend entirely on the way your information is organized (see below).

In this area the principle of searching in reverse chronological order becomes all the more important. An author who has done previous work in your field or one close to it may have forgotten or considered it unimportant to give a back reference to a compound. But for his basic approach and for the ideas and methods he employed—for these he has to supply ample justification which in most cases can be supported only by quoting prior work to the fullest extent. Otherwise, as he knows only too well, both referees and other workers in the field will be breathing heavily down his neck.

Naturally in a general search as defined in the heading of this section the number of potential sources of information is very large. What the beginner needs is a guide to the best and most useful of these, and to the scope and limitations of each. The following discussion is based on one man's view and thus necessarily subjective, but I believe it is based on somewhat greater than average experience.

Chemical Abstracts

By rights this should come first also in a general search, but in
practice, in the majority of cases, it has to be left as a last resort.
The reason is that either you will have great difficulty in finding
anything at all in the Index on the concept you are looking for,
or else you will be faced with an *embarras de richesse*. For example,
under the heading 'Ring Closure' the 1967–71 *Subject Index*
lists no fewer than 2300 entries in 36 columns. The advent, since
1971, of the *General Subject Index* has made the task a bit simpler.
There is more sub-division and there are new and additional
concepts (e.g. Ring Contraction, Ring Cleavage) which were
as good as absent in previous issues, but it is still very much a
case of trying to find a needle in a haystack. If you value your
eyesight and stamina do not use *Chemical Abstracts* for this kind
of literature search unless you have exhausted every other
possibility.

Theilheimer's *Methods of Synthetic Organic Chemistry*[12]

It is almost certain that on closer acquaintance with this thorough
and well-organized compendium on the synthesis and transformations of all classes of organic compounds you will become
addicted to it. This makes all the more unfortunate its horrendous
price (the latest volume costs one-fifteenth of its weight in gold).
A supplementary volume appears almost every year and the
literature coverage and choice of examples, while occasionally
on the subjective side, leave no room for complaint. Every
example gives structures of starting materials, intermediates and
products and brief details, with references not only for the
example in question but also for other relevant work. The most
important part of the work is the detailed and carefully cross-
referenced *Index*. I personally have yet to find one instance where
a reference for 'A, starting material for B' was not matched by
the same one under: 'B, from A'. The Index refers to the
example number and not to the page. It also lists reagents (inorganic

and metal–organic always under the metal or non-organic element), name reactions and reaction types. There is a special system and symbolism for division into reaction classes which looks rather intimidating and may frighten off the beginner until he realizes that mastering it is not at all essential for getting the full benefit out of this work.

At the end of each volume there is a Table listing supplementary references for examples given in previous volumes. This must be used with some care. For example: 'Vol. 1...23, Vol. 28...10' means that *page* 10 in Volume 28 contains an example or a reference relevant to *example* No. 23 in Volume 1. Each volume also has an introduction which sums up special developments since the appearance of the previous one.

At a rather late stage the publishers have realized that for the price of this work some improvement in the binding might not be inappropriate.

Houben–Weyl's *Methoden der Organischen Chemie* (4th Edition)[13]

After you have deciphered the fine gold lettering on the back, guessed which of five or six volumes might be the right one, admired a full-page photograph of a very important person, skipped through 2–4 pages of miscellaneous Forewords and Prefaces, waded through 14 pages or more listing journals most of which you never have encountered or never will encounter, and noted 2 pages of abbreviations employed, then—at last—will you come upon a meaningful List of Contents. With most parts of this monumental work (48 volumes so far) the initial effort will prove worthwhile. The first volumes deal with general laboratory practice and with analytical, physical and general chemical methods. They are a little out of date but should not be ignored on any account. The only recent addition to this group is a volume on *Oxidation* (and only the second of two). All the others deal with organic compounds: according to elementary content ('*Oxygen Compounds*', '*Nitrogen Compounds*'), functional groups (*Ketones, Nitro Compounds*), ring size (*3- and*

4-membered rings), special fields (*Photochemistry*), or special compound groups (*Macromolecules, Peptide synthesis, Organometallics*).

It is evident that the organization of *Houben–Weyl* has been rather haphazard over the years, and this together with the fact that many different authors have been involved lend an uneven quality to the work. Fortunately the List of Contents (once it has been located) is very helpful especially in conjunction with the Indexes. More recent volumes contain a welcome organization of information in tabular form.

In view of the enormous amount of information amassed in the work as a whole it is a great pity that it is so difficult to correlate it between its different parts. The emphasis is strongly on preparation and synthesis of specific functional groups, and much less on their interaction and on properties and transformations of compounds. Also, the organization of some parts appears to be quite arbitrary, for example the volume on *Oxidation*, which has been subdivided according to oxidation reagents (mostly by metallic element)—without any alternative classification or even cross-referencing according to substrates and products. As a result a lot of effort and time may be spent on collecting bits and pieces of information from widely different volumes. Worse still is the energy often consumed in finding out that a certain topic *cannot* be found anywhere. And yet: out of the blue, so to say, one may come across a proper review such as on the Diels–Alder reaction; and excellent as it is one cannot but wonder at the sudden whim that put it there in this general context.

I suppose that one of these days the editorial board of *Houben–Weyl* will get around to publishing some kind of General Index. Until then a frequent consulter of this work is advised to contrive some cross-referencing system of his own.

Technique of Organic Chemistry[14]

This is (or, rather, was) another rambling collection of volumes issued in the course of a number of years and by a number of different authors. These are indeed on various experimental techniques and topics, but the emphasis in most of them is rather heavily on their theoretical basis; and whatever they contain in really practical and useful detail is pretty much out of date for the contemporary practical organic chemist. In fact, in this regard it is difficult to think of a single volume in this series which has not been superseded by a more modern if not far better book or review. In short, the main reason why these volumes are mentioned here is that most libraries have them and many authors still quote them.

Since about 1970 the place of this series has been taken by another entitled, more simply, *Techniques of Chemistry*[15] (same Editor). Some of the volumes that have been published under that general designation are indeed useful and relevant for the experimental organic chemist. Among these one should mention the ones on *Organic Solvents* (Vol. 2, 1970), *Electroorganic Synthesis* (Vol. 5, 1974) and *Contemporary Liquid Chromatography* (Vol. 9, 1976). Also important are the three volumes on *Elucidation of Organic Structures by Physical and Chemical Methods*, which incidentally contain a chapter on asymmetric synthesis—for reasons that to me at least look quite unconvincing.

Organic Reactions[16]

This is the standard series collecting reviews on 'Name' reactions and others of a well-defined character. Each review has sections discussing mechanism, scope and limitations of the reaction discussed, but the really important parts for the literature searcher are the Tabular Surveys. Only a study of these can provide an answer to the cardinal question: how a specific reaction is compatible with a particular molecular environment and/or with the presence of other functional groups. Most of these Tables have had a lot of work put into them and this occasionally borders on the incredible, such as in the two-author review on the *Aldol Condensation* (Vol. 16), listing a total of 2359 references. All the same one must keep in mind that complete coverage is humanly impossible; every experienced searcher will find significant omissions in the course of time. In order to follow on from the date of the last review a good practice is to note which authors have been most active most recently in the field, and then to apply the Author Search approach (see below).

Other Sources of Inspiration

Fieser and Fieser's *Reagents for Organic Synthesis*.[17] This work, six volumes of which have appeared, delivers far more than its title might suggest, especially if you *read* it and not merely use it as a book of reference. It is a veritable treasure trove of reactions and transformations, by 'reagent association'. The 'Index of Reagents according to Types', included in every volume, is in fact an Index to reactions and methods, but the most valuable information will usually be picked up by just browsing with no definite object in mind. The only cautionary note to sound is in connection with the occasional and unexpected changes in no-menclature encountered from one volume to the next. But this is a very minor blemish on what I am sure is the best single invest-ment an experimental organic chemist can acquire at the present time.

Annual Reports in Organic Synthesis[18] and *Compendium of Organic Synthetic Methods*.[19] These periodical issues could be considered a sort of 'Pauper's Theilheimer', but their coverage does not go nearly that far. In the former the information is organized more by concepts ('Carbon–carbon bond formation') while in the latter it goes strictly by preparation and interconversion of func-tional groups. An advantage of both series is the way the infor-mation is displayed for instant recognition by clear structural formulae.

Organic Reactions in Steroid Chemistry.[20] The notion that steroids are somehow innately different from other classes of organic compounds is largely a myth—but a useful one. It would hardly be feasible to write a review on topics such as the intro-dunction of an alkyl group or a C=C bond, on hydrogenation or on the reduction of ketones, which would be valid for organic chemistry as a whole. But within the more restricted field of steroids such a task is not only practicable but also much to the point. Steroids have served for most types of reactions as a testing ground (mechanism, reagents, steric influences, optimum conditions, yields, etc.) and the results are often the best point of departure for application to other systems. Herein lies the significance of this work which follows on from an earlier book[21] and review[22] on the same lines.

Modern Synthetic Reactions.[23] This features a thorough and pretty up-to-date discussion on *some* types of reactions of special interest; it is particularly valuable because of the large number of references quoted. *Organic Functional Group Preparations*[24] is a welcome successor to the earlier *Synthetic Organic Chemistry*[25] which was a 'Best Buy' in its day. Volumes II and III are especially useful because they contain information on certain compound types which is not available in collected form in *Houben–Weyl* or elsewhere.

Friedel–Crafts and Related Reactions[26]; and *Formation of C—C Bonds.*[27] Even though these two works are on more specialized topics, as opposed to the ones discussed so far, they must be mentioned because together they cover a very large field which is entered by practically every experimental organic chemist at some time. In the four volumes of the first work practically every type of electrophilic substitution reaction, mostly but not exclusively on aromatic and heterocyclic substrates, is discussed and demonstrated. A good many of the chapters are by the Editor himself and the general standard is high. The tabular surveys give conditions and yields in a way that makes comparison between alternative conditions and between similar substrates an easy matter. In the case of polyfunctional substrates and products one must be careful not to rely entirely on one Table only, because it is not always obvious under which function they may be listed. An important and useful feature is a Cumulative Index which covers the work as a whole.

In the second work, spotted easily because of its unusual format, the first volume lists (with a short discussion in each case) examples covering C—C bond formation where a carbon bearing a functional group is introduced (e.g. hydroxymethylation, aminomethylation, carboxylation); the second deals with the more simple cases of alkylation, alkenylation, arylation, etc. It pays to get used to the method of classification employed in this work: sub-division is according to the bond type of the carbon atoms formed in the product, and the functional nature of the substrate and of the reagent. Thus, examples of the alkylation of a β-keto ester with a Mannich base methiodide will be found under: alkylation giving rise to bond between two sp^3 carbon atoms, alkylation of activated carbon atoms, by means of amines or ammonium salts.

Organic Syntheses.[28] This well-known collection is left here till last for a good reason. The very fact that it contains detailed and painstakingly checked directions for preparing specific compounds or closely related ones (in itself performing a great service to organic chemists as a whole) can be a psychological trap for the beginning researcher. He is tempted to believe that directions for preparing aromatic ketone A by a Friedel–Crafts reaction can be applied verbatim to making another aromatic ketone B, or that the same tools and procedures described for a reaction on a molar scale can and should be used for one on a millimole scale. The result is nearly always frustration and discouragement. Fortunately the latest issues take account of this and tend to give procedures which can serve as a general model and which in many cases have not been published in any way elsewhere.

ON REVIEWS IN GENERAL

The problem is that there are so many of them, and hence the need first of all to look *for* a review, let alone the question of *how* to look in it. Fortunately there are indexes of reviews and recommended is the one sponsored by the Chemical Society.[29] The latest cumulative issue has appeared quite recently (1976).

A review is of course where someone should have done all the work for you. In practice this rarely turns out to be the case and it is necessary to understand why, by knowing something of the circumstances and terms of reference under which they are generally written. To this end one could divide reviews roughly into three categories.

In the first are those whose authors were required to cover a subject comprehensively up to the time of writing. Obvious examples are those found in *Organic Reactions*[16] (and it is to be hoped that the declared policy of updating previous chapters will be continued). Other series where such reviews can be found include *Chemical Reviews* (American Chemical Society). Obviously books and monographs on a reaction, method or field can and should be in this class.

The second category are those reviews in which coverage of a subject was done for a limited period of time. These can be found in series such as *Annual Reports* (Chemical Society), in the accompanying *Specialist Periodical Reports*, in *Organic Reaction Mechanisms*,[30] and in a large number of other series usually prefixed by '*Annual Reports* . . .', '*Progress in* . . .', '*Advances in* . . .', or '*Fortschritte* . . .'. In most of these coverage of the literature is likely to be especially thorough, at least for the period in question, because the authors knew of their assignment ahead of time and were thus prepared for a thorough and systematic scanning of the literature. For this reason careful study of such reviews combined with that of earlier references quoted will probably give the best overall picture of the present state of knowledge in any subject covered.

The third group are those reviews where the author had been working intensively in his field, considered that he had accumulated a good deal of literature data and thus felt compelled to offer a review for publication; or where he was considered by others to be leader in his chosen or self-originated field and was then invited to write a review with no strings attached. These are likely to put particular emphasis on the author's own contribution and views and on that previous work which he considered to have the most direct bearing on his own. They will often be found in series such as *Accounts of Chemical Research, Chemical Society Reviews* (formerly *Quarterly Reviews*), *Angewandte Chemie*, and in journals such as *Synthesis, Bull. Soc. Chim. France*, and *Tetrahedron*.

All in all, any review in which references to the author's own work exceed 10% of the total number should be treated with reserve.

SEARCHING BY AUTHOR

Reactions, methods and concepts usually become associated with the name of a leading personality. Hence it should be possible to obtain further information on the ideas he has initiated by following up his name once such an association has been established. The Author Index of *Chemical Abstracts* is the obvious place to look but first one has to decide *which* name to look out for. Until recently it was not clear from a publication authored by, say, Pickles, Gherkin and Chutney, who was the senior author and hence the source of inspiration. The situation became clearer when you discovered that there were 20 entries for Chutney and only one for each of the others; obviously Chutney was your man. This approach is not quite foolproof: Chutney may by now be over 60 and spending most of his time on committees or travelling from one meeting to the other while it may be Pickles who is now publishing (or perishing) in a junior academic position—quite likely he is at present following up and expanding the original idea as his very own.* Nowadays most journals require marking with a star or other sign the author 'to whom correspondence should be addressed'. In most cases this is indeed the senior author unless he does not want to be bothered by people asking for reprints.

The author search approach may quite frequently uncover information which will otherwise escape you, for example where, as occasionally happens, an article has been passed over by *Chemical Abstracts*. It may be the only feasible one when you have to find out what has happened to some idea in the recent past beyond the cut-off date of the last review on a subject. Hence it is good to

---

* Any resemblance to actual persons is of course purely coincidental, and in any case rather unlikely.

keep in mind that for a number of years now it has been developed on an impressive scale by the *Science Citation Index*. This appears four times a year in three main parts. The *Source Index* lists alphabetically the names of authors and other sources who have published in the period under review and gives the titles and location of their publications. The *Citation Index* lists alphabetically names of authors and their publications, and under each enters other authors and publications who have quoted these. The *Permuterm Index* lists 'subjects' (see below) and enters the names of authors who have published on them, indicating at the same time which have published more than once on the subject and which have published on it uniquely. A detailed explanation of the whole system is given by an article by E. Garfield.[31] It is clear that by concurrent consultation of all the three indexes one can quickly arrive at some overall picture of who is working on a particular idea (or, say, its 'Old Boy' network), and hence perhaps of its state of development. I say 'perhaps' because much important work does appear in less accessible journals.

One is also tempted to consider the *Permuterm Index* as a subject index in its own right. However, the 'subjects' are arrived at by computer manipulation of significant terms within the titles and subtitles only and thus depend entirely on which words the author of each article has seen fit to include in the title. For that reason this Index has its limitations and must be used with caution.

KEEPING UP TO DATE

Chemical Abstracts Keyword Index

Each weekly issue of this publication now contains this Index which alphabetically lists combination of 'keywords' which are taken both from the title of the publication and from the abstract. It is thus superior to the *Permuterm Index* mentioned above, and provides a quick but still not entirely reliable method of finding out if the current issue contains an item of interest.

Current Contents

This merely reproduces the content pages of current periodicals and provides a keyword index based on these. This, like the *Science Citation Index* mentioned previously and the following periodical, is published by the Institute for Scientific Information.

Current Abstracts of Chemistry and Index Chemicus

This is a weekly journal which summarizes current articles by a brief abstract and, more importantly, by flowsheets showing structural formulae of important products and intermediates. Each summary also indicates techniques used, spectral information quoted, and if the main aspect of the paper is the description of a new synthetic method, an X-ray crystallographic analysis, etc. It is a good idea to take a number of issues of this on a long train journey, its perusal demanding a more-than-average degree of concentration (not to mention eyesight). The reason is its format, and the necessity for sifting the grain from the chaff. Literature coverage is wide and also indiscriminate; it includes all the important journals admixed liberally with 'trivial information from obscure ones; and the time it takes to decide which parts to skip is not much more than that taken by glancing through the whole issue. Each issue also contains a Molecular Formula Index covering the compounds mentioned; and keyword, author and other specialized indexes. In spite of its shortcomings it is perhaps the only practicable alternative to reading the original literature itself.

In the end one cannot escape the conclusion that there is no better way of keeping up to date than to read each journal as it comes in. This is difficult even to contemplate, but one must always remember that the most significant information one may ever pick up can be that found coincidentally: in a casual reference, from a chance remark, and in a fortuitous experimental observation. A research worker who keeps this in mind will always have the edge over another who relies on the easy ways out.

ON ORGANIZING YOUR INFORMATION

'Having it at your fingertips' is the figure of speech commonly used. You can turn it into reality only in the form of a Card Index. There should be a card for every topic, and there should be cross-referencing cards pointing in all feasible directions. You should always have a supply of cards with you; envelopes or paper napkins get lost, mixed up or used for other purposes. There is no need to use the customary stiff cards; slips of ordinary paper cut to appropriate size will do just as well.

One could make suggestions on classification, such as dividing a Card Index into two parts: functional groups, and reactions, methods and concepts, each arranged alphabetically. But in the end you will have to devise your own system, and change and modify it as time goes on and as experience is accumulated. For this reason the system of using holed punchcards with selection by needle, while seeming attractive to start with, will not be adequate in the long run.

Your Card Index is an extension of your memory cells and as such should remain strictly your private domain. As your stack accumulates your popularity as a source of information with your co-workers will grow, but—never ever lend out cards to anyone else.

HINTS ON SCANNING A JOURNAL

Today's successful executive will find it essential to go through some kind of speed-reading course to cope with the daily heap of printed matter on his desk. Some similar training is needed by a searcher of the current scientific literature. For the synthetic organic chemist there is a particular need for some system or approach that will enable him to concentrate on the horse of facts, results and experimental detail rather than on the cart of exposition and justification in a paper.

With Communications (*Tetrahedron Lett.*, *J. Chem. Soc.*, *Chem. Commun.*, *Chem. Lett.*, *Synth. Commun.*, *Angew. Chem.*, and the back parts of *J. Am. Chem. Soc.* and *J. Org. Chem.*)

scanning the list of contents is usually enough to bring all the important facts to one's attention; full details will in most cases be absent, in any case. Besides, Communications are or should be brief by definition and skimming through them is not such a great effort.

Full papers need a quite different approach. You will probably turn your attention first to those journals devoted entirely to organic chemistry: *J. Org. Chem.*, *Tetrahedron*, *Synthesis*, *J. Med. Chem.*, *J. Heterocycl. Chem.*, *J. Chem. Soc. Perkin Trans.*, *Annalen*, and the relevant separate issues of *Bull. Soc. Chim.* and *Acta Chem. Scand.* Other journals (*Helv. Chim. Acta*, *Chem. Ber.*, *Coll. Czech Chem. Comm.*, *Bull Chem. Soc. Jpn*) group all papers in organic chemistry together in the contents list but unfortunately not in the journal itself. A third group (*J. Am. Chem. Soc.*, *Canad. J. Chem.*, *Austral. J. Chem.*) make no attempt to help the reader by any kind of sub-division.

With full papers it simply will not do to rely on the title of the paper or on the author list when making a choice of which to read. Any fundamentally novel finding will in all probability have been published already in Communication form. Most importantly: a paper on purines may contain details on the preparation of a useful β-dicarbonyl intermediate; one on the mechanism of a Claisen-type rearrangement may give valuable information on an important organometallic reagent; and one on carbohydrates on how to open a sterically hindered epoxide by malonate ester. All such potentially valuable information will be lost to you if you decide *a priori* that purines, Claisen-type rearrangements or carbohydrates have nothing to do with your own line of work; and it may equally escape your attention if you decide to rely on a Keyword Index.

Hence the advisability of going through each issue of a journal from cover to cover. 'Going through' should mean looking for those items likely to be most helpful in serving as some sort of anchor; and first and foremost there are the structural formulae, reaction flow charts and schematic indication of reagents and conditions used. If every author would also keep this in mind his work would receive much greater attention.

Often this is not easy because of an ever-increasing tendency to list compounds under a basic structural formula with variations indicated as R, R', R''; a, b, c; $n = 1$, 2, 3 etc.; this is not the fault of the author but of the editor intent on saving money.

After having fixed those items in your mind you will be able to plan mentally the scanning of the Experimental Part. In this it is safe to assume that any author who has taken care that his work is reliable and reproducible will have taken equal pride in describing the work accurately, clearly and unambiguously. For example, the word 'treat' ('the solution was treated with . . .') is frequently a convenient term to use when the author has forgotten exactly how it was done. Opening statements like 'General directions as described in Part X' may hide the fact that they are inapplicable in many cases in Part XI. Apart from this General Directions and Methods often contain much valuable information on techniques used and references to preparation or purification of starting materials, solvents and reagents which should not be missed and are only too easily overlooked.

Only then should you skim the main text of the paper for anything which might be additional information and reference of value rather than padding and reasoning after the event. Should you encounter a statement like 'the presently described method is greatly superior to previously reported ones' your first reaction should be not to look at the high-yield last step but at the overall yield and effort of the steps leading up to it from an easily available starting material, or how much effort and expense go into the special reagent used. Unfortunately quite a number of recent 'elegant' innovations are in that category.

Statements such as 'Further work is continuing' or 'Additional results will be reported in a subsequent paper', once an implied threat and 'Keep Off' notice (especially in the German literature), are now rarely more than an off-hand promise honoured in the breach rather than in the observance.

2
On Carrying Out
Small-scale Reactions

The object of this chapter is to discuss some aspects of conducting a reaction on a 1–20 mmol scale. This is the order of magnitude most frequently encountered by a preparative organic chemist of advanced student or post-doctoral level, and where choosing the right tool for the job can be of critical importance. Reactions on a scale much smaller than this are usually too specialized for any general discussion; and for anyone working on a scale larger than this there is no lack of textbooks at various levels of excellence, and even electronic and audiovisual aids for those who can afford them, to serve as a guide.

CHOOSING THE RIGHT TYPE OF REACTION FLASK

Most organic preparative reactions involve two or more operations such as stirring, heating or cooling, temperature control, addition of reagents and so forth; and to enable such operations to be performed at the same time two- or three-necked flasks with adapters and other accessories are usually employed.

Before going into this in detail one should perhaps make a few observations on an important but rather neglected subject: the shape of the flask. Figures 1, 2 and 3 illustrate three alternatives. The first shows that beloved of so many laboratory supply houses and university departments who have somehow never thought out the basic problems involved in running a chemical

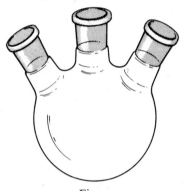

Fig. 1

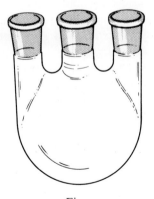

Fig. 2

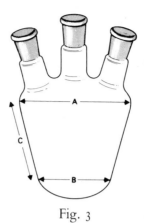

Fig. 3

reaction (typical remark heard at a venerable British university: 'we've 'ad those for forty years and they've always served us well'.). If you have ever tried to use this type of flask properly you will no doubt have noticed (a) that the thermometer is either broken by the stirrer or vice versa, or else will measure nothing except the temperature inside the empty stirring vortex, (b) that the angle of the side arms from vertical—anything up to 45°—is just right to ensure a good chance of breakage under the load of the condenser or addition funnel plus contents which they hold, and (c) that this angle makes clamp support of such ancillary items most awkward, and glass being what it is it will never prevent breakage anyway.

Figure 2 shows a better but more expensive design. However, rarely is the side arm properly aligned to allow ingress of the thermometer without interfering with a stirrer blade, and often the condenser, thermometer and addition funnel will interfere with the stirring motor and rod unless the latter is sufficiently long or magnetic stirring is used. Also, when this type of flask is of less than 250 ml capacity the gap between the necks becomes so narrow that clamp support of the center neck is impossible.

The best design is that shown in Fig. 3. It is also at present the most inaccessible from commercial sources. Such flasks were sold at one time (without standard joints) by Sovirel under the name 'sulfonation flask', perhaps because in this type of reaction the problems mentioned above are particularly evident. However, it seems they ceased producing them just at a time when a number of laboratories were waking up to the advantages of this shape. They are still listed by Normalschliff Glasgeräte, for example, but only from a 100 ml size upwards and with an unnecessarily large center joint. You will find it well worth it to have a glassblower make a number of these. Suggested relative dimensions are $A : B : C = 1 : 0.71 : 0.85$. The two-necked 25 ml and 50 ml versions will prove to be the most useful in small-scale work—in these dimension A is $c.$ 37 mm and 46 mm respectively. The shallow bottom curve is ideal for trouble-free and steady magnetic stirring and the straight sides allow for maximum effective immersion of a thermometer and inlet tube. Moreover the wide top makes it possible to fit even a 50 ml flask with four side necks without causing mutual interference.

In most reactions temperature control is involved and hence it is logical to allow for this from the outset by fitting a standard screw inlet with cap to fit a normal thermometer instead of one of the standard joint side necks (Fig. 4). In this a tight seal is ensured by an inner Neoprene® ring; naturally it can also be used for a gas inlet tube and when not in use it can be closed by inserting a Neoprene® or Teflon® washer. Alternatively such inlets are obtainable with a standard joint (Fig. 5).

For large-scale reactions the two-part 'resin' flask (Fig. 6) has many advantages. The cover portion can incorporate up to four standard joints with minimum mutual interference and with such a flask, working up messy products such as from a Friedel–Crafts reaction is very much easier. Certain suppliers list apparatus for small-scale reactions which are constructed on the same principle, but in these the various parts are very critically aligned and thus tricky to repair and the large spherical joint connecting the two halves is either liable to freeze or, if greased, may introduce a good deal of contamination into the product.

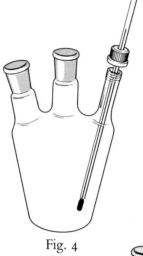

Fig. 4

Fig. 5

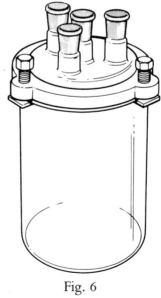

Fig. 6

DRYING TUBES

In most laboratories one encounters the bent type as illustrated in Fig. 7. This will prevent contamination by the drying agent but it includes too much unnecessary internal surface and will often break at the critical moment. In a small-scale set-up, and particularly when working at room temperature or lower, the simple straight type (Fig. 8) is more compact and durable and makes it possible to use standard joints at both ends. One can carry the concept of compactness further, in the form of a combined drying tube and thermometer inlet (Fig. 9) or a drying tube-come-cold finger condenser (Fig. 10).

Fig. 7

Fig. 8

As for drying agents the best one in the majority of cases is silica gel of the self-indicating type previously heated to 150 °C in a high vacuum and stored in a rubber-stoppered bottle. Where basic reactants and solvents are used, especially reactions with liquid ammonia, there is probably no alternative to the use of calcium oxide or barium oxide although I have yet to come across a supplier who sells these in uniform particles of suitable size. Incidentally, tubes containing calcium oxide always seem to crack while absorbing moisture and they should be tightly stoppered at both ends when not in use. Calcium sulfate, a universally applicable desiccant, is now available in the right kind of size (e.g. Fluka's Sikkon®) but its capacity is limited and it is difficult to tell when it is exhausted.

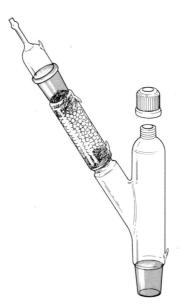

Fig. 9

For temporary drying purposes, such as stoppering a flask containing a solution dried by azeotroping off moisture, cotton wool is handy and effective but this must be dry to start with and a supply should be kept in readiness inside an 80 °C oven.

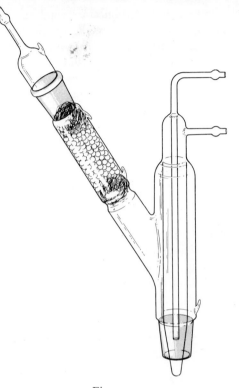

Fig. 10

THERMOMETERS

There is a well-established extension of Murphy's Law which states that the part of the scale which matters is the one hidden inside the stopper. If only more thermometers were made (and ordered) whose scale begins well above the bulb and which are corrected for a reasonable (say 1 cm) immersion instead of the usual 7–10 cm!

Other sins committed by manufacturers and compounded by meek and uncomplaining research workers: scales which become invisible after short use due to leaching-out of the pigment— rectification of this by using boot polish is at best a messy and always a temporary solution; and bulbs of ridiculously large size, a special problem in small-scale work and in distillation. In buying thermometers and most other kinds of laboratory hardware 'economy' definitely does not pay.

MAGNETIC STIRRING

This is one of those 'trivial' topics where the correct choice can make all the difference. Of the four generally used types of stirring bars as illustrated (Figs. 11(a)–(d)) the third one (of octagonal cross-section) usually gives best results. The first one is cheapest and it is guaranteed to behave erratically. The (American ?) 'football' type (Fig. 11(d)) is fine so long as you are willing to face the increased risk of breakage with this hefty version.

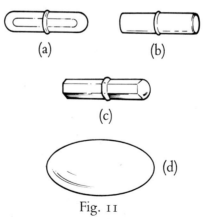

Fig. 11

It has been reported[32] that placing a sheet of aluminum between the stirring motor and reaction flask improves evenness of stirring, and I can vouch from personal experience that this works even (to the consternation of my electrical engineering friends) when the motor is all aluminum-enclosed to start with. Another phenomenon which should be taken into account is that there is usually some specific optimum height above the motor for effective steady stirring and this should be allowed for when setting up a reaction.

No material other than Teflon® is generally suitable as covering for magnetic stirring bars. A glass covering is always under strain unless known for certain to be perfectly annealed and hence it will crack in the majority of cases; polyethylene and polypropylene will swell if not actually disintegrate in the majority of hot organic solvents. Teflon® is indeed attacked with blackening by solutions of alkali metals in liquid ammonia and such like. However, here is a method (discovered by chance after working-up an alkaline epoxidation) for cleaning such bars: in a hood, cover the bar with 30% hydrogen peroxide, add an equal volume of 20% sodium hydroxide solution, wait for the reaction to subside and complete by heating on the steam bath. If necessary, repeat. However, when enough of the Teflon® coating has been destroyed for the magnet to show through, such a bar should be discarded.

HEATING WITH STIRRING

Most people tend to use a combined stirrer-hotplate with a water or oil heating bath. This may be convenient but in small-scale work it usually means a lot of wasted heat and little latitude for raising the reaction flask to an optimum stirring height (see above). Heating mantles may seem even more handy and a lot cleaner in use besides, but the majority of them do not allow a magnetic stirring field to penetrate and there is a greater risk of overheating unless they are of the shallow type which are not available any more. Perhaps the best and at the same time cheapest solution to this problem lies in the use of a miniature immersion heater (Fig. 12) such as can be bought at any hardware or electrical store for quickly heating a glass of water or for

Fig. 12

sterilizing baby bottles. This can be connected to a variable resistance; the lower part of the heating element should be flat and not bent at right angles.

For temperatures up to 100 °C, water covered by a thin layer of mineral oil (to prevent evaporation) should be used; for higher temperatures a good list of various heating media is given in *The Chemist's Companion*.[33] It is always good policy especially in distillation to place a magnetic stirring bar in the heating bath, this incidentally will help along the one inside the flask.

COOLING

Very few suppliers offer Dewar-type vessels of a shallow type suitable for conducting small-scale reactions at low temperature. Any that can be found at all are in a limited range of sizes and expensive. Above all, their protective metal enclosure makes magnetic stirring well-nigh impossible. A suitable substitute lies in the use of two nestling shallow glass evaporating dishes, spaced for insulation by glass or cotton wool, whose shape follows that of a round-bottomed flask and thus leads to economy in the amount of cooling medium required (Fig. 13).

On the latter subject *The Chemist's Companion*[34] is once again a highly recommended source of information.

Fig. 13

STOPPERS

Here is another, by no means trivial, but often worrying subject. Standard joint glass stoppers are cleanest but do not provide a perfect seal unless greased and that once again means contamination especially in small-scale work. Also they offer no resistance to internal pressure and if a reaction should suddenly 'take off' a glass stopper can be a dangerous projectile. Rubber or, better still, Neoprene® stoppers are of course better in this respect, but they will not stand up to hot organic solvents for longer periods of time. This is particularly noticeable with serum stoppers (see later).

In many cases, particularly when a reaction is conducted at room temperature or below, polyethylene or polypropylene stoppers are quite satisfactory and provide an excellent seal. A very good range of these in sizes to fit various standard joint openings are sold, for example by Protective Closures Inc. under the name of 'Caplugs'.

The best solution, of course, is to have Teflon® stoppers machined by a workshop, but they make quite a hole in one's budget. Otherwise using glass stoppers with Teflon® sleeves is an acceptable alternative.

BOILING AIDS

On a small scale in particular, wooden boiling sticks (applicators) are quite out and should be thrown away or put to some quite different use. They are messy and easily absorb their own weight of solution. With boiling stones you always take a chance. Many sold are ineffective and some are by no means inert (one British firm I could name had the nerve to sell some which turned out to be more or less pure calcium carbonate). Carborundum chips are best but, as usually sold, they are not of uniform mesh size and should be graded by sieving to obtain the average size suitable for the scale at hand. Being dark, they can usually be separated from crystalline material, and that should forestall the usual temptation of including their weight in the measured yield. However, they scratch glassware and for prolonged periods of heating under reflux, especially in combination with stirring, their use is ill-advised.

On the whole it is best to rely on stirring alone, magnetic or otherwise, even in a distillation. A capillary inlet is fragile and difficult to adjust; and at high temperatures argon instead of air must be introduced to prevent oxidative decomposition. The cleanest boiling aid is to insert a melting point capillary, with the closed end up, and this should be used when recrystallizing for analysis. However, remember that if and when heating is interrupted even for a second the capillary will fill up with liquid and will then cease to function.

DRYING APPARATUS, FLAMING OUT

Oven drying of apparatus, especially in small-scale work, is never good enough. By the time you have assembled the apparatus it may well have adsorbed enough moisture to spoil a reaction on a millimole scale—remember that 1 mmol of water is only 18 mg. This means of course that keeping the internal surface to a minimum should be uppermost in your mind when choosing your glassware. It also signifies that the only sure way to have dry apparatus is to flame it out in a stream of dry inert gas just before starting a reaction, and this too has to be done properly. Start flaming just below the drying tube (do not heat the drying agent itself—this will only desorb traces of moisture which are perfectly retained at room temperature) and work your way slowly to the farthest exit. The apparatus should then be closed, allowed to cool to room temperature, and then the flask cooled to -40 °C, all under positive dry inert gas pressure. If there is any marked evidence of condensation the whole process will have to be repeated.

ENSURING AN INERT REACTION ATMOSPHERE

It is difficult to think of an organic reaction which does not benefit by being run under nitrogen or, better still, argon, unless aerial oxidation of the product is intended deliberately. In many laboratories this problem is tackled by an array of often colorful and sometimes grotesquely (not to say downright Rabelaisian) shaped rubber balloons at strategic positions. This looks cheerful on color slides, but one soon finds out that this approach has severe limitations. The nitrogen trap described by Johnson and Schneider[35] is much better, and a logical extension of the principle of this apparatus is shown in Fig. 14. A stout-walled capillary tube (a) of inner diameter of c. 2 mm and of 85 cm length is fused at the bottom into a shallow mercury reservoir (b) provided with an outlet. At the top it is connected via the safety bulb (c) to a manifold tube having between three and seven outlets (d) with stopcocks and ending in a three-way stopcock (e) for either applying vacuum or admitting inert gas. All the stopcocks used must be of highest quality and carefully greased and preferably fitted with a retaining spring. Since this apparatus is all in one piece the wooden board on which it is mounted must be reinforced with a steel rod (otherwise warping will crack the assembly). The outlets should be connected to inert gas and apparatus by thick latex rubber tubing which is probably less permeable than the ordinary red type (among the common elastomers 'Saran' appears to be best in this regard but flexible tubing made of this may not easily be available).

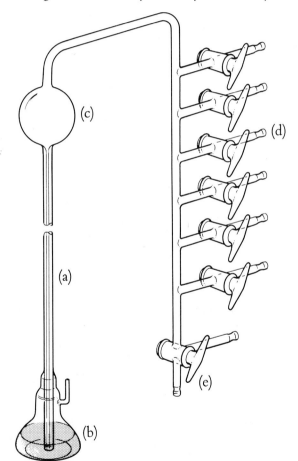

Fig. 14

This apparatus makes it possible (a) to carry out several reactions under inert gas at the same time, (b) to obviate the need for equilibrated glassware such as addition funnels, and (c) to store temporarily a sensitive solution, reaction mixture or product while preparing another reaction set-up in the vicinity.

THE RIGHT REACTION SET UP

This is a most important topic in small-scale work, but unfortunately it is difficult to give categorical answers to general questions such as whether in principle to use a three-necked flask or a two-necked one with side arm adapters. Much depends on the type of reaction, the conditions, the number and type of operations involved, and of course not least on the interfitting glassware available. Here are a few general rules which could perhaps serve as guide:

1. The reaction flask should not be more than one-third full. This is to guard against 'eventualities' (foaming, sudden exothermic reactions) and also to make provision for possible working up of the reaction mixture in the flask itself. Of course this does not apply where in the course of the reaction the volume is decreased by distillation.

2. There should be a minimum exposure of greased joints. This can only be ensured by careful planning on the basis of glassware available. For example, where reflux is involved the cooling surface should be as low as possible such as by using a cold finger condenser (see Fig. 10). In fact there should be no greasing where it is not essential to keep air and moisture out.

3. While taking the above points into account the total internal surface should be kept to a minimum, such as by using the smallest size condenser commensurate with the amount of cooling required, or the smallest addition funnel to hold the amount of addend involved.

4. Dropwise addition of a reagent should be directly into the reaction mixture and not down the sides of the flask. Otherwise a precipitate which may form cannot eventually be dislodged and may occlude one of the reactants.

5. Top-heaviness must be avoided, and the amount of clamping kept to a minimum. Glassware parts should be firmly held together by springs and fused-on hooks as far as possible. Hooks should be fused on near every standard joint, and when doing so one should stop and think every time whether the hook should turn 'up' or 'down'. The forked clamps often employed take up too much space.

6. Magnetic stirring should be used whenever possible. If the stirring bar should get stuck you should try another motor or a different stirring height. Mechanical stirring in small-scale work should be considered only when there is no alternative, such as with a highly viscous and heterogeneous mixture.

7. All apparatus should be set up on a movable platform ('Jack') and not the bench itself. This will make it easier to control heating and cooling.

8. With any kind of set-up 'branching out' should be in a vertical rather than horizontal direction. For instance, when distilling off and in distillation in general why not use a Friedrichs condenser or an 120° adapter allowing for downward distillation and thus save lots of valuable space, instead of doing it the way nearly all practical textbooks thoughtlessly illustrate it? Admittedly one can go to the other extreme: at institutions in certain countries research is reported to be done mainly in a vertical direction, using ladders going up the ceiling. Perhaps this is one way to try to increase a laboratory's 'output norm'.

SUPPORT STANDS AND FRAMEWORKS

Perhaps this is the right point for a few remarks on this subject and on bench organization in general. Support stands should be kept in the corner and used only where there is no other alternative. Their base plates always get in the way of some item or other, and apparatus such as stirring motors, hotplates and steam baths exhibit the perverse habit of just failing to sit properly on the normally sized base plate, especially when mounted on feet or studs. A framework, firmly mounted on the bench and constructed from aluminum rods and connectors is always preferable, but its construction must be planned with some thought beforehand. The greater part of it should allow for maximum free movement and leeway in a vertical rather than a horizontal direction and at least up to 50 cm above the bench. Many commercial ready-made frameworks do not allow for this, being constructed throughout with standard square openings which suit only vacuum lines used by physical chemists. Part of the framework should have exposed lengths of rod pointing up (for holding cork support rings) and others horizontally (for rolls of cleaning tissue and aluminum foil). At the periphery, arrangements should be made for holding items of equipment for instant availability such as plastic beakers to hold capillary pipettes, glass rods and spatulae. The best overall arrangement is usually discovered by trial and error in the course of time and this should be taken into account when putting up a framework for the first time.

HOODS

As will be stressed further on, much of your work should be done here (or in the 'fume cupboard' if you are in the U.K.), if only for reasons of safety and health; and here rational bench organization is particularly important. All permanent or semi-permanent items such as nitrogen traps, manometers or guard vessels should be mounted close to the wall and going up vertically rather than sideways. The inevitable problem of corrosion in a hood can be met only by having most if not all metal fittings (including screws, of course) made of stainless steel. It does not take long to find out that the extra cost will be worth it.

DRYING OF SOLVENTS AND THEIR TRANSFER

The accompanying Table 1, based on the author's experience, is meant to reconcile some of the conflicting advice encountered on the subject of solvent drying and purification; and it takes into account the comparatively high state of purity now shown by many commercially supplied solvents even when delivered in bulk and described as being of 'practical' or 'technical' grade quality.

A few points, often not stressed sufficiently, must be made. First of all, most hydrocarbon and halocarbon solvents can be dried as thoroughly as by any other means by azeotropic water removal—distilling off between 5 and 25%—and thus drying not only the solvent but also the containing vessel. It follows that in any reaction where anhydrous conditions are essential the inclusion of an inert hydrocarbon solvent such as benzene as at least part of the system should be considered. This can be done, for example, by placing a benzene solution of one of the reactants in the flask, boiling off part of the benzene in a current of dry nitrogen (always remembering the considerable toxicity of this solvent !), and then adding other dry solvents and reactants. This can probably be done in the majority of reactions where the use of solvents such as tetrahydrofuran or dimethylformamide alone is specified.

The use of extruded sodium wire for drying solvents is in my humble but considered opinion a practice that is as dangerous and unnecessary as it is widespread. Its drying capacity is limited to start with and ceases with the formation of the adhering layer of sodium hydroxide; and the disposal of voluminous strands of partly decomposed sodium poses special problems and is the cause of a goodly proportion of laboratory fires that I have heard of. Perhaps it is a good idea to send all sodium presses to the departmental museum or present them as a gift to whoever can find some other use for them.

Storage over molecular sieves is now routinely advocated and is probably most generally effective. However, it is often not realized that they must be freshly activated, usually by heating to 300–350 °C (metal bath) in a high vacuum. This should be done immediately before use, or the activated material should be stored under totally anhydrous conditions. It must be done even with a 'freshly opened' new bottle because many suppliers do not bother to use the proper kind of container for this sort of material. The process of activation usually produces fine dust which of course cannot be sieved out at that stage and in many solvents may take some time to settle.

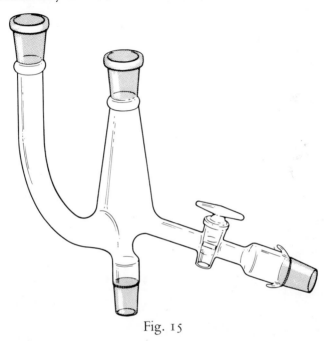

Fig. 15

Now to the special problems posed by solvents such as tetrahydrofuran, 1,2-dimethoxyethane and dioxan. First of all, peroxides must be tested for and removed, by methods described in the literature, and also mentioned in Table 1; and in this connection you should be warned that the potassium iodide–starch paper strips often used cannot absolutely be relied upon. After this they are best pre-dried by distillation from calcium hydride, using a reflux–take-off adapter such as illustrated in Fig. 15 (in which the stopcock should have as wide a bore as possible), but for many purposes this is not good enough. On a larger scale, where 20–500 ml of dry solvent is required at one time, further drying can be effected by distillation from

TABLE 1

Purification, drying and storage of common solvents

Solvent	Boiling point (acceptable range)/°C	Preliminary purification	Further drying and purification	Recommended storage
Pentane Hexane Cyclohexane Other Alkanes	36 (2–3) 69 (2, 5[a]) 80.7 (1)	Wash several times with conc. sulfuric acid to remove olefins if necessary, wash water, dry $CaCl_2$, collect after wet forerun	Rarely necessary; if so by azeotropic water removal	Up to 500 ml in glass-stoppered bottles; above that and for long periods in dark screw-capped bottles. No sense in keeping over molecular sieves
Benzene[b] Toluene[b] Xylenes	80.1 (0.5) 110.6 (1) 144.5 (ortho) 139 (meta) 138.3 (para) (1)	Dry over $CaCl_2$, refractionate, reject first 5–10% of wet forerun	Boil off from or redistil into reaction flask, rejecting first 5% (see text)	in dark screw-capped bottles. No sense in keeping over molecular sieves
Dichloromethane Chloroform Carbon tetrachloride 1,2-Dichloroethane	40 (1) 61.2 (0.5) 76.8 (0.5) 83.5 (1)	Wash with water, dry over $CaCl_2$, redistil, collect after 5% wet forerun	Redistil from P_2O_5; on small scale and in special cases pass through alumina (basic, act. I) directly into reaction flask	As above, but chloroform for longer periods in tightly closed full bottles and in darkness
Diethylether Diisopropylether	34.5 (1) 68.5 (1)	Test for peroxide; if positive wash with 5% metabisulfite soln, then with sat. NaCl; dry over $CaCl_2$; distil (but not over conc. H_2SO_4)	Small amounts: pass through up to 10%/wt alumina (basic act. I); with larger amounts best to use absol. diethyl ether from cans	Best in cool, dark place in nearly full screw-lidded metal cans. For long periods seal with 'parafilm'[c]

	b.p. (m.p.)	Predrying	Purification	Storage
Tetrahydrofuran 1,2-Dimethoxy-ethane (Glyme)	65.5 (0.5) 84 (1)[d]	Stand overnight over KOH, decant; test for peroxide; if positive stir with up to 0.4%/wt NaBH$_4$, overnight, add CaH$_2$, fractionate but not to dryness	Distil from potassium under argon (see text), small amounts pass directly into flamed-out reaction flask through alumina (basic act. I)	In dry plastic-insert screw-capped 100 ml bottles over basic active alumina under argon. For long periods seal with 'parafilm'; Dioxan best kept frozen in refrigerator
Dioxan	101.5 (1)[d] (m.p. 11–12)		Distil from sodium under argon (see text)	
Carbon disulfide	46.5 (1)	Redistil in hood from small amount of phosphorus pentoxide; use only water bath heated by steam	Shake with small amount of mercury, redistil from phosphorus pentoxide	Don't! Avoid leaving around laboratory
Ethylacetate Methylacetate	77.1 (0.5) 57 (1)	Dry over active CaSO$_4$ (Sikkon) and/or anhydr. K$_2$CO$_3$, decant, distil carefully	Fractionate from up to 5%/wt of acetic anhydride	Over activated molecular sieves 5A in tightly closed bottles
Other esters boiling below 100 °C			Refractionate	
Acetonitrile	81.5 (0.5)[d]	Predry over MgSO$_4$,[e] then over anhydr. K$_2$CO$_3$, decant, distil from CaH$_2$	Fractionate from P$_2$O$_5$; Small amounts: pass through alumina (basic, act. I) directly into reaction vessel	Over activated molecular sieves 3A, best in 100 ml dated bottles

TABLE 1 (cont.)

Solvent	Boiling point (acceptable range)/°C	Preliminary purification	Further drying and purification	Recommended storage
Acetone	56.2 (0.5)	Distil over 2 °C range, dry over anhydr. CaSO₄, decant, redistil	If used for oxidation reactions reflux over sufficient KMnO₄ to retain violet color, distil, dry, fractionate. Very pure via NaI addition compound	Over freshly activated molecular sieves 3A
2-Butanone	79.5 (0.5)	Fractionate off water azeotrope (b.p. 73.5 °C), dry this and remainder separately as for acetone		Over freshly activated molecular sieves 5A
Methanol	64.5 (0.5)	Simple fractionation now usually sufficient even with bulk grade		
Ethanol	78.3 (0.5)	From 'rectified spirit' (95%) most economically by reflux and distillation from CaO (at least 1.5 times amount needed to bind water present)	Predried material redistilled from CaH₂ best directly into reaction vessel	In small bottles over freshly activated molecular sieves 3A
Isopropanol	82.5 (0.5)	Fractionate, collect after water azeotrope (b.p. 80.3 °C), dry latter as for ethanol		
n-Propanol and higher alcohols	97.2 (0.5)	Fractionate, collect after aqueous azeotrope forerun		

t-Butanol	82.5 (0.5) (m.p. 25.8)	Water azeotrope b.p. 79.9 °C. Treat as for isopropanol	As for previous alcohols but care needed in distillation—solid may block condenser!	As for previous alcohols; bottles best kept in warm place during cool season to save bother of 'thawing out'
Ethylene glycol, higher glycols	198, 68–70/4, 108–110/28, (2)	Fractionate in vacuo, collect after 5–10% forerun. High latent heat of vaporization!	Refractionate after dissolving up to 1% weight of sodium	Best in 100 ml plastic insert screw-capped bottles (very hygroscopic!), larger amounts only over large excess of molecular sieves
Nitromethane Nitroethane	101.3 (1)e 115	Dry over $CaCl_2$, decant, fractionate	Refractionate from molecular sieves 4A	Over molecular sieves 4A
Formic acid	101 (1) (m.p. 8.3)	Fractionate, best under somewhat reduced pressure. Can be dried further by reflux and distillation from phthalic anhydride. Water azeotrope has b.p. 107 °C (22.5% water)	On prepurified material freeze completely, allow to thaw to extent of 10–20%, decant (all while protected from moisture), use remainder	In screw-capped bottles
Acetic acid	118 (0.5) (m.p. 16.6)	Refractionate after adding up to 5% acetic anhydride and up to 2% CrO_3		
Pyridine Methyl pyridines	115.5 (0.5)	If very crude dry over KOH, decant, fractionate	Reflux with CaO, BaO or very active basic alumina, refractionate	In tightly closed dated bottles over molecular sieves 5A

TABLE I (cont.)

Solvent	Boiling point (acceptable range)/°C	Preliminary purification	Further drying and purification	Recommended storage
N,N-Dimethyl formamide[f] N,N-Dimethyl acetamide N-Methyl pyrrolidone	153, 42/10, 55/20 (t); 166, 58–59/11, 63/18 (t); 202, 78–79/10, 96–97/24 (t)		Stir overnight with CaO, BaO or alumina (basic act. I), then refractionate in vacuo	Best over freshly activated molecular sieves in small dated bottles. With larger amounts (above 500 ml) amount of sieves should be large to take up moisture introduced on frequent opening
Dimethyl sulfoxide	190, 50/3, 72/12, 84–85/22 (t) (m.p. 18.5)	Fractionate in vacuo, rejecting first and last 10%. Avoid distilling at atmospheric pressure	Stir with calcium hydride overnight, then fractionate from calcium hydride in vacuo. Can be further purified by partial freezing if dry	
Hexamethyl phosphoric triamide	235, 68–70/1, 115/15, 126/30 (t) (m.p. 7)		Stir 1 h with CaH₂ at 100 °C under reduced pressure, then refractionate in vacuo	In small (50 ml) plastic insert screw-capped bottles under argon and over activated molecular sieves 13X or over oil-free NaH if available

[a] Cheaper 'hexane fraction'.
[b] Assumed free of sulfur compounds such as thiophene.
[c] Can be stabilized or restabilized by adding up to 0.001% of a dihydric phenol.
[d] May frequently be supplied in state of purity inferior to that claimed; purification calls for extra care.
[e] May be encountered in part as low-boiling azeotrope with water.
[f] Reported to be light-sensitive; probably best kept in dark bottles at all times.

lithium aluminum hydride (although one keeps hearing of unexpected explosions when using this reagent for that purpose), or from sodium in the case of dioxan, or, with the appropriate amount of caution, from potassium in the case of tetrahydrofuran and dimethoxyethane (the melting point is just below the boiling point of the former solvent and sodium is useless for these two solvents). The apparatus used must of course be perfectly dry to start with and should allow for maintaining an inert atmosphere throughout. Two suitable designs are shown in Figs 16 and 17:

In the first, designed by Dr J. Schreiber (E.T.H., Zürich), either reflux or outflow can be effected by tilting the adapter (A) connected by the large spherical joint (B) upwards or downwards respectively. This arrangement allows for flushing a reaction flask attached at (G) with solvent, by returning it into the system by both upward tilt and rotation of adapter (A). It should be noted that point (p) has to be some distance (say 10 mm) above point (r).

The second design has the advantage that a measured amount of solvent can be dispensed. The condensate can enter a graduated receiver (B), made from a large graduated centrifuge tube, from which, via a double oblique three-way Teflon® stopcock it can either return into the system or be allowed to run into a flask attached at (G). Both pieces of apparatus described can be attached permanently to the multi-outlet nitrogen trap (see Fig. 14).

On a smaller scale (0.5–15 ml solvent) all effort put into drying may well be nullified by the relatively large amount of moisture introduced during actual transfer even when a pipette is used. In such a case it is necessary to adopt an arrangement whereby the measured amount of solvent is transferred directly into the flamed-out reaction vessel against a stream of dry nitrogen or argon; and this is illustrated in Fig. 18. A burette or small cylindrical graduated funnel with a capillary outlet tube and containing a layer of sand and plug of glass wool is dried for at least 1 h at above 100 °C, and a column of the requisite amount[36] of high-quality basic alumina (act. I) is added. Through this the solvent is allowed to percolate as illustrated. This also allows for adding more dry solvent at a later stage since ingress of moisture via the capillary drip tip is practically nil. This method should always be used for diethyl ether. Predrying of this solvent presents special and obvious problems; it is best to rely on the commercially available sealed metal cans the contents of which are sufficiently dry for, for example, Grignard reactions, provided transfer is by pipette only.

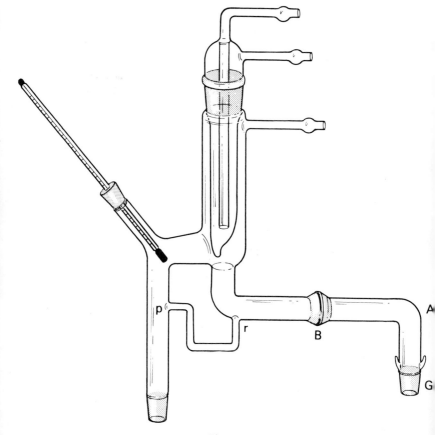

Fig. 16

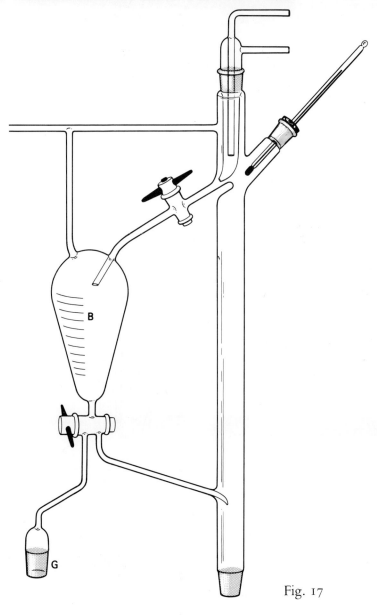

Fig. 17

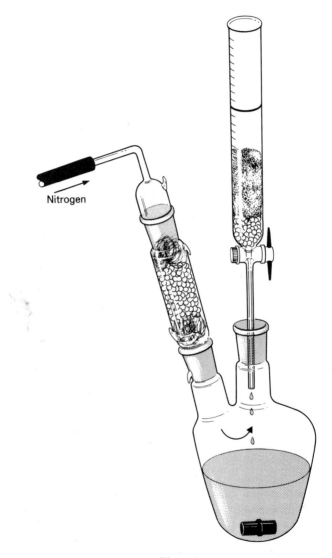

Nitrogen

Fig. 18

TRANSFER AND ADDITION OF SENSITIVE LIQUID REAGENTS AND SOLUTIONS

Figures 19(a)–(c) illustrate and describe the use of a special type of burette which has proved itself for dosage and addition of alkyl lithium solutions and other sensitive liquids. A standard burette (any size from 5 to 50 ml) is fitted with a Teflon® stopcock and a capillary stem long enough (say 15 cm) to reach to the bottom of a 250 ml bottle. The top of the burette is modified into an adapter tube for attachment of rubber tubing or of a bulb-type three-outlet pipette filler.

Immediately after a storage bottle is opened it is flushed with nitrogen by inserting a small bent tube connected to the multi-outlet nitrogen trap (Fig. 14); the stem of the burette is introduced a short distance and flushed with nitrogen and evacuated several times. After this the liquid is drawn up to the desired volume (Fig. 19(b)). After closing the stopcock (and of course the storage bottle) the top of the burette is attached to a second nitrogen outlet. Part of the contents can be used for titration, such as by the method of Watson and Eastham[37] after which the stem is inserted into a one-holed stopper and attached to the reaction flask (connected to a third nitrogen outlet) for dropwise addition (Fig. 19(c)).

When the product of a reaction has to be transferred to another vessel (e.g. addition funnel) or onto another reagent (as in carbonation of a Grignard or lithium reagent) the method shown in Fig. 20 is both simple and effective. A two-holed rubber stopper containing a twice-bent tube is inserted into the flask. The tube is lightly lubricated for facile raising and lowering; it should be long enough to reach the flask bottom and may start in a widened portion with a plug of glass wool for filtration. The second hole in the stopper is to put one's finger on for instant flow control. Before the actual transfer under inert gas pressure (all parts should be spring-connected) nitrogen flushing can be done while the tube bottom is above liquid level. When the tube is finally lowered the flask should be inclined so as to get the magnetic stirring bar out of the way.

58

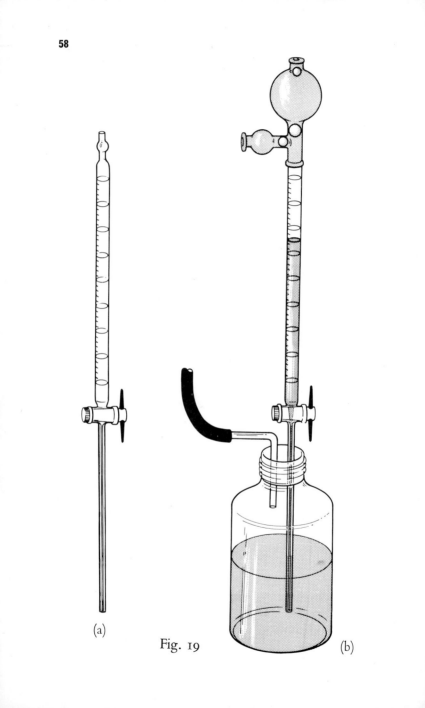

(a)

Fig. 19

(b)

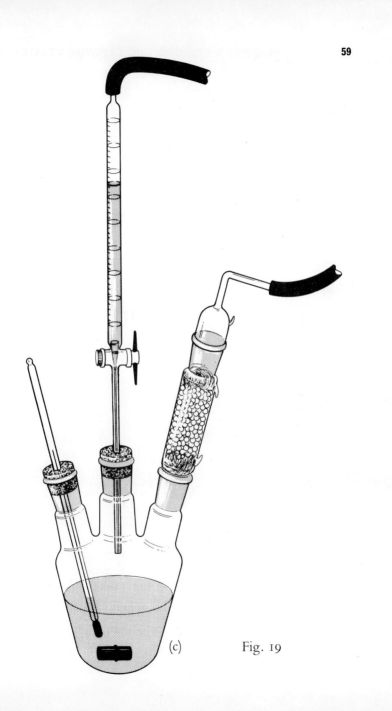

(c)   Fig. 19

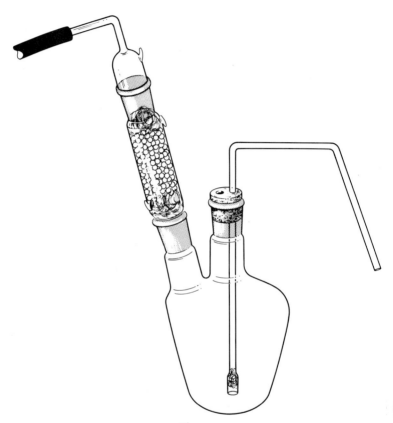

Fig. 20

When one wants to transfer a sensitive solution from a large storage bottle to smaller containers an arrangement using a three-holed stopper (Fig. 21(a)) is used analogously; and in the case of storage bottles with too narrow a neck to hold such a stopper the combination shown by Fig. 21(b) can be employed. In this the delivery tube can slide freely, held by a rubber sleeve, inside a T-tube and this contains a small hole, as indicated by the arrow, for finger-tip control.

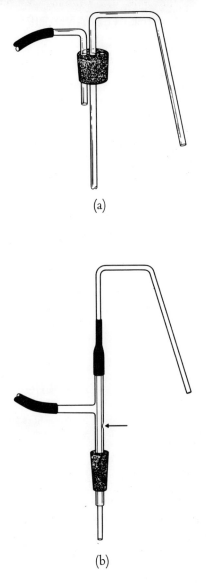

(a)

(b)

Fig. 21

GRADUATED PIPETTES AND PIPETTE FILLERS

You should always keep a plentiful supply of pipettes because none should ever be used more than once in between cleaning and drying. Those graduated to the tip and of the 'blow-out' type are to be preferred, if only to eliminate the temptation of returning unused liquid. However, few if any manufacturers have yet come to the logical conclusion of starting the graduation of these from zero at the tip!

Pipette fillers of the rubber bulb three-outlet type as shown in Fig. 19(b) are recommended. With these a large volume range can be withdrawn and dispensed. In fact they can be used for withdrawing a volume larger than the nominal bulb capacity because the bulb can be again evacuated by squeezing while still attached to the pipette or burette, something that cannot be done with the piston type of filler. Unfortunately they vary a great deal in quality. Of a batch obtained from a certain West European supplier a good proportion were useless to start with (faulty valves) and most of the remainder deteriorated after a short time. In our experience the 'Propipette' model (Franz Bergmann K.G.) has given best results.

ON RUBBER SEPTA AND SYRINGES

Some people swear by their use. Many others swear while using them. Only the most expensive types of septa are made of an elastomer which is resistant both to chemical attack *and* to the results of repeated puncturing. Also it is difficult to find septa in a sufficient range of sizes to fit various openings or even just to fit all standard joint openings in common use.

And as for syringes: they are expensive and need little provocation for breakage—as witness all those firms advertizing to have you send them in for repair. And now to needles: somehow they turn out to be either too short or too long or too wide or too narrow for the purpose at hand; their connection to the barrel is not always as tight as it should be and may break, or otherwise detach itself just when least expected. Filling a large syringe with a liquid of some viscosity may take a long time and can be hard work. Dropwise addition from a syringe, except when using an expensive and space-consuming device, can never be as simple and accurate as from a burette. In my opinion the use of a syringe can be justified only where volumes of less than 0.05 ml are involved.

Having got these subjective remarks off my chest and already hearing the howls of dissent in the distance (and not only from syringe manufacturers who are doing very well, thank you very much), I hasten to add that they were made mainly in connection with preparative work. Far be it from me to suggest an alternative to the use of syringes and septa in, say, gas chromatography.

WORKING WITH SODIUM AND POTASSIUM HYDRIDES

Sodium hydride is now generally available as a dispersion in mineral oil or grease containing between 50 and 80% of the hydride by weight. This is never quite homogeneous and hence whatever concentration is stated on the label must be taken as correct only within $\pm 5\%$. In most reactions some excess is employed anyway. The mineral oil must be removed in the reaction flask by washing out under an inert atmosphere, and the best way to do this is as follows. The weighed amount is placed in the flamed-out flask while dry inert gas is passed through the system; it is then covered by hexane and the suspension is stirred for several minutes. The hydride is then allowed to settle while the flask is inclined to one side, after which it is gently inclined in the opposite direction (see Fig. 22; drying tube and other attachments not shown). The supernatant washings can then be removed almost completely by capillary pipette. This process is repeated until a drop of washings placed on the ground-glass joint will evaporate in the inert gas stream without leaving a trace; four to five times is usually enough. The hydride is then washed once again with the dry

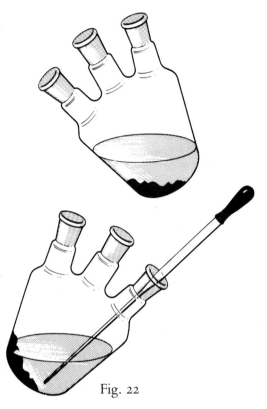

Fig. 22

solvent used in the reaction, preferably added through an active alumina column (Fig. 18).

With potassium hydride still greater care must be taken and this is also more difficult to handle. Its mineral oil suspension settles very easily and weighing it is quite out of the question. It *may* come in plastic bottles with an internal plastic lid (see below on Ordering alkali metals and hydrides, p. 156). If so, the lid should be punctured with a hypodermic needle connected to dry argon, after which it is barely prized open and the well-agitated suspension withdrawn using a pipette whose tip has been cut to give a 2 mm diameter opening (a polypropylene pipette will do very well for this). The volume withdrawn should be on the basis of the stated concentration of the suspension and on its assumed specific gravity of 1.15. The contents of the pipette are then washed into the flask by rinsing with dry hexane from above, and the material is washed free of mineral oil as described for sodium hydride. The puncture hole in the storage bottle lid should be sealed with paraffin wax or with 'Parafilm'.

ON THE USE OF DIPOLAR APROTIC SOLVENTS

There is no doubt that these have greatly advanced experimental organic chemistry; and the accompanying Table 2 summarizes some useful information on the most important of these; much of this is available only in brochures and other commercial literature.

The chief drawback of these solvents is that it is often so difficult to get rid of them once they have served their purpose. This applies particularly to hexamethyl phosphoric triamide (HMPTA). Practically every month sees the publication of yet another example of how various reactions proceed under milder conditions, or in better yield, or proceed at all, when using this or a similar solvent, but on somewhat closer examination one will often find that the substrates have been carefully chosen so as to make isolation of the product a simple matter; it is non-polar and hence isolable by extraction with a hydrocarbon solvent such as pentane; not to speak of cases where on looking just a little closer one discovers that the product was not actually isolated at all but that results and yields were arrived at on the basis of gas-chromatographic or spectral analysis alone. When one tries to apply the method or conditions to more complex and polyfunctional compounds one usually finds that isolation of a pure product is a lot more difficult.

With all these solvents it is worthwhile investigating first whether the use of the very minimum amount together with a more easily removable co-solvent (ether, benzene, tetrahydro-furan) will not give just as good results. Often there is no need

Dipolar aprotic solvents—some useful data

	N,N-Dimethyl formamide	N,N-Dimethyl acetamide	N-Methyl pyrrolidone	Dimethyl sulfoxide	Hexamethyl phosphoric triamide
Boiling points/°C (mm Hg)	153(760), 90(100), 42(10), 22(3)	166(760), 25(2), 63(18)	202(760), 78–79(10)	189(760), 126(100), 74(10), 44(2)	233(760), 120(16), 99(6), 66(0.5)
Melting point/°C	−61	−20	−24.5	18.6	7
Dielectric constant	36.7	37.8	32.3	48.9	30
Basicity, as $\Delta\delta^\infty$ (CHCl$_3$)[a]	1.30			1.34	2.03
Some selected solubilities	(in g/100 ml at 25 °C): AgCl > 16.5, CuCl$_2$.2H$_2$O > 14.7, CuSO$_4$ 1.7, KCN 0.2, KCNO 0.11, KCNS 17, K$_3$Fe(CN)$_6$ <0.05, KI > 23.6, KMnO$_4$ > 16.5		More than 10% at 25 °C: (NH$_4$)$_2$S, Pb(OAc)$_2$, PbCl$_2$, S, KMnO$_4$, KCNS, ZnCl$_2$. KF: 3% at 190–200 °C	(in g/100 ml at 25 °C): KI 20, KNO$_3$ 10, KNO$_2$ 2, AgNO$_3$ 130, NaI 30, NaNO$_3$ 20, NaNO$_2$ 20, ZnCl$_2$ 30, CuI 1, CuBr$_2$ 1, LiBr 31.4, LiNO$_3$ 10, LiClO$_4$ 31.5	(in g/100 ml at 25 °C): NH$_4$Cl 4.4, AgNO$_3$ 33.3, CuSO$_4$ 13.7, NaCl 0.78, KCl 0.2, NaNO$_3$ 8.8, Na$_2$SO$_4$ 0.1

[a] $\Delta\delta^\infty = \delta^\infty - \delta$, where δ^∞ is chemical shift of chloroform in the solvent at infinite dilution and δ is that of chloroform in an inert solvent (cyclohexane).

Commercial sources of information: N,N-Dimethylformamide and N,N-dimethylacetamide—E. I. du Pont de Nemours and Co., Inc.; N-Methyl pyrrolidone—GAF Corporation; Dimethyl sulfoxide—Crown Zellerbach, Chemical Products Division; Hexamethyl phosphoric triamide: Pierrefitte-Auby.

Other sources: General—A. J. Parker, The use of dipolar aprotic solvents in organic chemistry, *Adv. Org. Chem.* **5**, 1 (1965); Dimethyl sulfoxide—D. Martin and H. G. Hauthal, *Dimethyl Sulphoxide*, van Nostrand Reinhold, Wokingham, 1971; HMPTA—H. Normant, Hexamethyl phosphoramide, *Angew. Chem. Int. Ed. Engl.* **6**, 1046 (1967).

even to use enough to give a homogeneous mixture. With HMPTA in many cases no more than one molar equivalent is necessary. When the product of the reaction is acidic this solvent can be removed almost quantitatively by extraction from the alkaline solution with a chlorinated solvent such as chloroform;[38] and where it is stable to acid conditions it can be simply destroyed by hydrolysis. In any event the best policy when working up reaction mixtures containing dipolar aprotic solvents is to use only hydrocarbons (benzene, if not hexane) for product extraction.

ON THE ADVANTAGES OF USING REAGENTS IN SOLUTION

Many sensitive reagents can be bought in solid form. As examples one could cite strong bases such as sodium methoxide, potassium *t*-butoxide or lithium diisopropylamide. However, a solid has to be weighed on a balance; and unless this and its subsequent transfer are done in a drybox (of which more later) it will inevitably spoil even if this is done 'very rapidly'. On the other hand, liquids and solutions can always be measured accurately and transferred without such disadvantages and since most reactions are run in solution anyway one should always think of preparing, keeping and using such reagents in solutions of known concentration. For instance, sodium methoxide: it is not difficult to make a solution of up to 4M concentration of this frequently used base and to transfer this to and keep it in plastic insert bottles (see below under Bottling, p. 153) in which it can be kept unchanged for a long time. Such a solution can be titrated very accurately; and where the use of this base is specified as a suspension in, for example, benzene this can be achieved by removing the methanol as its azeotrope with the non-polar solvent. The same goes for alkali *t*-butoxides although the maximum achievable concentration of these is much lower and also the shelf life of such solutions is shorter.

Obviously, the principle of using standardized solutions of reagents can apply also when these themselves are liquids. For example, alkyl aluminum compounds are very dangerous to work with in a neat state, titanium tetrachloride fumes so much that you cannot see what is happening when it is transferred or added without dilution, and methoxy- or ethoxyacetylene are so volatile that you cannot be sure whether half the amount is not blown away in the nitrogen stream against which it has to be transferred. Particularly in small-scale work it is advisable in such cases to prepare a standard solution once and for all (for example, in the above cases in toluene, carbon tetrachloride and pentane respectively). This is not only more convenient and less dangerous but it also makes it easy to dispense small quantities accurately.

When you yourself make up such solutions you know, of course, the concentration. On the other hand, where such solutions are sold commercially the supplier is liable to be, shall we say, cagey if not downright evasive on this point. This is of course particularly objectionable in case of a reagent whose titration is not an easy matter. More will be said on this further on (see p. 160).

DRYBOXES AND GLOVEBOXES—THEIR USE AND MISUSE

These still seem to be much more 'in' in American laboratories, possibly as a legacy of the Manhattan project. It is necessary to make a clear distinction not only in name but in function. A drybox should be used *only* for storage, weighing and transfer of sensitive solids (not liquids) and preferably for no other purpose. A good all-purpose and reasonably sensitive balance, best of the top-loading type, should be a permanent fixture inside and it is unfortunate that so few commercially available ones are suitable as far as overall dimensions are concerned. Also there should be an arrangement of easy-to-get-at shelves—a two-tiered turntable ('Lazy Susan') is ideal. Other permanent inside features should include pencil and paper and a reliable clock-type hygrometer (not one of the 'wet bulb' type, of course). For most purposes wide dishes containing activated self-indicating silica gel and calcium oxide will ensure a sufficiently dry and CO_2-free atmosphere.

The main trouble with dryboxes is that they are usually common equipment. They will never remain both 'dry' and clean unless under constant eagle-eyed supervision or in the ideal environment where tidiness and order are imbibed and become part of the national character. Your guess is as good as mine as to the central European country where such dryboxes can still be seen.

A glovebox is for performing all kinds of operations in an inert atmosphere most of which, if you really gave some thought to it, could be performed more easily by the right kind of bench set-up. *A propos* this subject: who has not had the automatic urge to scratch some part of the anatomy while both hands are inside the gloves? It can be as messy as one likes provided the atmosphere inside is really inert and that is much more difficult to achieve than generally realized. Several hours' purging by pure argon may be required to pass the acid test: an electric light bulb without the glass glowing for more than a few minutes before burning up.

On reading through the (usually U.S.) literature one sometimes sees the use of gloveboxes carried to ludicrous extremes. For instance: 'The suspension was filtered in a glovebox under pressure.' Try that if you can in your glovebox, and see what happens.

The disposable transparent glove bags now available are, needless to say, cheaper by one or two orders of magnitude.

OTHER BOOKS ON THE SUBJECT

The foregoing remarks were meant to apply to reactions where the use of sensitive reagents and the associated special techniques and precautions are encountered at only a few points in an operation sequence. This is the case in the majority of work described in the modern literature; and there it is best to work and plan on the premise that no organic chemist has more than two hands.

The situation will be different when you will be doing work in which a whole sequence of operations has to be performed in an inert atmosphere, in a vacuum or in a closed system. These may include gas transfer, filtration, evaporation, distillation, transfer to closed cells for spectroscopic examination and the like. In such a case you can probably do no better than consult the book by Shriver[39] on the subject. This contains a wealth of information not only on techniques and apparatus but also on materials, methods of working them and their sources of supply, tables of physical constants and methods of gas purification. In addition, a recent work on organoboranes[40] contains an appendix on techniques, methods and apparatus found to be most useful in that special field; it also describes methods of analysis and for automatic gas generation which have quite general applicability. Other useful sources of information of this nature are the brochures issued by the Ethyl Corporation[41] on the handling of organoaluminum compounds.

Finally, I would highly recommend the reading of Brandsma's book,[42] *Preparative Acetylenic Chemistry*. Even if you happen not to work in this particular field you will find that the methods described in detail by someone of great experience who has done all the work with his own hands will constitute a good general guide for working with sensitive compounds and will in any event demonstrate that planning ahead every detail, dismissed by so many others as unimportant trivia, can spell the difference between success and failure in experimental work.

3
On Isolating and Purifying the Product

CAPILLARY PIPETTES

To begin with, a few words on these, among the most important and yet simplest of all your tools in small-scale preparative work. They can be straight, slightly curved for reaching into awkwardly shaped vessels, or into whatever shape necessary, for example, for withdrawing a sample from a Kugelrohr tube (Fig. 23), and are to be used for working up (Fig. 24), for filtration on a very small scale through cotton or glass wool (Fig. 25) and in general wherever transfer of a liquid, solution or crystallization mother liquor is required.

Most of the commercially available 'disposable' ones, supplied in gross lots, are too thin and of insufficient volume for the Neoprene® nipples generally available. The best sizes are those drawn from tubing of 8–9 mm outer and 6–7 mm inner diameter, and you had best draw your own and always have a

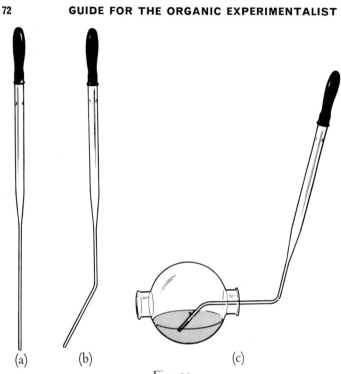

(a)   (b)   (c)

Fig. 23

plentiful supply available. They do not have to be thrown away after use; in most cases washing with water and ethanol is sufficient and if not there will always be room for a couple more in the next chromic acid bath. Besides, you are bound to develop a personal attachment to several of them.

WORKING UP

This is where probably the greatest number of avoidable mistakes are made by the beginning researcher. The reason is not only the half-hearted if not sloppy way the subject is taught the beginning student but also the manner it is airily glossed over in the literature by standard phrases such as 'the customary

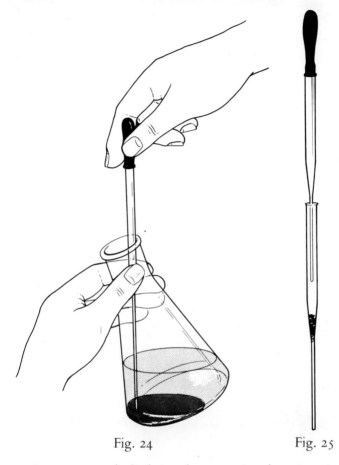

Fig. 24   Fig. 25

working up . . .', 'isolation by extraction (or trituration) gave . . .', or 'chromatography of the crude reaction product . . .'. Another case in point is:

'The Mixture was Poured into Water'

You should always approach with skepticism anything written in an Experimental Part about the final state of a reaction, more often than not by someone who did not actually do the work himself. For instance, how often does one encounter the above statement in connection with a Friedel–Crafts reaction or a Claisen-type condensation, only to find that this is ludicrously impossible!

Rule Number One

Whenever a reaction set-up is disconnected for work-up the first thing to do is to remove grease from joints. This is best done by wiping them several times with a tissue paper lightly soaked in a solvent such as carbon tetrachloride; for narrow openings a pair of tweezers should be employed. It is difficult to over-emphasize this point; no solvent should ever be added for extraction before this has been done. Moreover, I have yet to come across any water-based detergent that will remove silicon grease completely—manufacturers' claims notwithstanding. This type of grease has many advantages over the hydrocarbon types such as the Apiezons, but it has a pernicious habit of creeping into one's product, as witness the characteristic broad band at $1050–1100$ cm^{-1} in the infrared spectrum of so many beginners' preparations. It may also find its way into the glassware itself—a point which experienced glassblowers never tire of making.

Rule Number Two

Before extraction, water-soluble solvents used in the reaction such as methanol, ethanol or tetrahydrofuran should be removed as much as possible at the rotary evaporator. Failure to do so is one of the most frequent reasons for low yields. With high-boiling solvents (ethylene glycol, dimethyl formamide, dimethyl sulfoxide) this of course is not such an easy matter, but even then it may well be worth the trouble of getting rid of as much as possible in a high vacuum to begin with.

Proper Choice of Extraction Solvent and Vessel

Choosing the right solvent is no problem when the product and its properties are well known, but that is not so in the majority of cases and it may always be advisable to examine it first on a microtest-tube scale. The fact that it separates as an oil does not mean that a convenient low-boiling but mediocre solvent such as diethyl ether can confidently be used. All too often it is on adding this that at least part of the product may crystallize out and cannot be redissolved. If the first extraction is attempted in the original round-bottom flask used for the reaction then at least there is no problem in removing the unsuitable solvent at the rotary evaporator and trying another one.

If at all possible, solvents heavier than water should be used for extraction. An instructive demonstration for any beginner is to show him two conical separating funnels or round-bottom flasks, each with 100 ml water, and then add 50 ml of ether to one and 50 ml of dichloromethane to the other. Greater economy is not the only factor. You will be surprised to find how the extractibility of compounds normally soluble with difficulty in chloroform (e.g. dicarboxylic acids) improves dramatically after adding up to 10% of tetrahydrofuran. This 'synergistic' phenomenon is not uncommon, and mixtures of solvents (ether–benzene, chloroform–ethyl acetate) are often better than the sum of their constituents. When using such mixtures one must naturally ensure that the proportions used will lead to a mixture which is definitely either lighter or heavier than water.

A frequent mistake is to add too much water and/or too much extraction solvent, and then end up having to use a 500 ml separating funnel for working up 1 mmol of product. The amount of water should be minimal but must take into account the solubility of the inorganic salts finally present. For example, when sodium dihydrogen phosphate is used for acidifying a reaction mixture the salts produced have a low solubility particularly when lithium ion is present, and this must be thought of in advance.

The points made so far mean that the working-up process should be an integral part of the planning of a small-scale reaction, and that it should be done as far as possible in the reaction flask itself. The use of separating funnels should be avoided altogether and a capillary pipette used instead. This takes more time, but is cleaner, more accurate and prevents contamination. Washing with acid or alkali, further extraction of layers and final drying should be done using a consecutive series of Erlenmeyer flasks.

Snags Frequently Encountered

1. *Some carboxylic acids and enols may form insoluble salts.* When this is discovered (usually too late) it can be disastrous for a systematic and quantitative working-up process, more so because the resulting mess is often difficult to filter. Here again the importance of preliminary micro-scale examination is indicated. The solution of the problem, if any, usually lies in a change of cation (from sodium to potassium or lithium and occasionally ammonium). If this is of no avail it will be best to separate the bulk of the acidic/enolic material from the total crude product by crystallization, and then working-up the mother liquor.

2. *The acidic or enolic product is not extracted sufficiently by aqueous alkali.* This happens with tertiary and hindered carboxylic acids and with hindered phenols. The solution lies in lowering as much as possible the polarity of the organic phase, by substitution (benzene for ether) or adding hexane as much as possible without precipitating or oiling out one of the constituents; this will change the distribution in favor of the aqueous alkaline phase. The same principle obviously applies to the extraction of a weakly basic amine with dilute mineral acids.

3. *Aqueous alkaline solutions of organic acids (or acidic solutions of organic bases) tend to dissolve neutral organic material.* The latter of course includes solvents, but that is not so much of a problem as the dissolution and carrying along of a neutral reaction product. Hence, successive alkaline extracts of an acid (or acidic extracts of an amine) should each be separately back-extracted at least twice to be on the safe side, and the washings added to the neutral portion.

4. *Emulsions.* This common problem can be associated with any one or more of the above. The trouble is that every case is different. What should be tried are: (a) filtration of the whole under *moderate* suction through a reasonably wide bed of Celite or Filtercel—the culprit may be a minute amount of amorphous solid derived from an impurity in one of the reagents, (b) cautiously dissolving a neutral electrolyte (sodium chloride, sodium sulfate) in the aqueous phase, (c) adding a more polar solvent (ether, ethyl acetate or even a small amount of methanol) to the organic phase, and, most important of all, (d) patience and waiting.

Other Important Points

Solvents such as diethyl ether and ethyl acetate are never dried completely by either sodium sulfate or magnesium sulfate. In such cases a higher-boiling or azeotroping solvent should be added before final solvent removal to remove water completely. Carbon tetrachloride is best for this purpose, and especially so when, as is usually the case, the crude product is to be examined by infrared or NMR spectroscopy. While on this subject: one should make it a general rule to take infrared spectra in solution and thus get a reliably comparable picture. Taking them in Nujol mulls or KBr discs should be done only where solubility problems arise or when there are other really convincing reasons for this.

Filtration of the dried product solution should be into a flask with an opaque and not clear (KPV) joint. Complete transfer of the product by washing through is then clearly indicated by allowing a drop of the filtrate to evaporate on the joint. However, if anything shows up with 'pure' ether it means only one thing—the presence of ether peroxide.

If the yield of your product leaves much to be desired even at the crude stage, try salting out the aqueous layers: with ammonium sulfate for acidic and with sodium chloride for neutral and basic products, and then extract again. Very often you will be agreeably amazed. In fact simple extraction after intensive salting-out should usually be done before the use of a continuous extraction apparatus is contemplated—and every experienced worker knows how difficult it is to find one that 'happens' to be of just the right size.

There are of course always cases where an organic product is so water-soluble that complete evaporation of the solution followed by selective extraction with an organic solvent is the only course left open. Here it is not always realized that many inorganic salts can be appreciably soluble in some organic solvents, and the following Table 3, based on data given in the literature,[43] gives some semi-quantitative and comparative figures to enable one to choose the appropriate conditions in many cases.

Primary purification of a product to remove highly polar, resinous and colored impurities is always advisable. Often Kugelrohr distillation (see below, p. 102) can be used. With neutral products percolation in benzene through a short adsorbent column can be most effective (for this it is best to use Florisil rather than alumina or silica), and often just adding some adsorbent after addition of the drying agent will take care of the worst contamination.

SOLVENT REMOVAL

The rotary evaporator has by now become standard equipment and rightly so. Together with the apparatus itself the manufacturer will probably supply a mongrel-shaped flask which you are invited to use for all your evaporations large and small. No doubt you will find some use for it eventually. What he does not supply and should have is a suitable adapter to guard against splashing and sudden ebullition which may lead to loss, and to contamination of both the product and of the apparatus. A suitable one is illustrated in Fig. 26. In this the center tube should be straight and not curved so that if any liquid does splash up it can be returned into the flask by capillary pipette. If your rotary evaporator is of the home-made type cooling

TABLE 3

Solubility of inorganic salts in organic solvents—comparative data (weight loss per hour on Soxhlet extraction with refluxing solvent)

Salt	Methanol	96% Ethanol	abs. Ethanol	Acetone	Ethyl acetate
Na_2CO_3	0.3 g	10 mg	4 mg	0	0
$NaHCO_3$	0.8 g	15 mg	15 mg	0	0
K_2CO_3	4.3 g	50 mg	50 mg	0	0
Na_2SO_4	5 mg	0	0	0	0
K_2SO_4	0	0	0	0	0
$NaCl$	1.2 g	0.15 g	70 mg	0	0
KCl	0.7 g	0.1 g	70 mg	0	0
$NaBr$	vs	c. 7 g		30 mg	0
NaI	vs	vs	vs	vs	2.3 g
KI	vs	vs	vs	3.3 g	10 mg
KNO_3	0.35 g	0.2 g	50 mg	9–12 mg	0
$KClO_3$	0.17 g	70 mg	30 mg	6–11 mg	0
$KClO_4$	0.2 g	70 mg	40 mg	0.25 g	6 mg
NH_4Cl	s	0.5 g	1.9 g	0	0
H_3BO_3	vs	vs	vs	c. 2 g	0.1–0.2 g

vs, very soluble; s, soluble.
No weight loss of any of above with: benzene, petroleum ether, carbon tetrachloride, carbon disulfide, chloroform, or diethyl ether (except with latter, H_3BO_3: 0.15 g).

Fig. 26

should be arranged to be by means of a large cold trap (containing ice or dry ice–isopropanol) rather than a condenser, this not only leads to a better vacuum but also recovers solvents much more efficiently and such a set-up can be connected to high vacuum with much less risk of implosion. A simple way of raising and lowering the whole assembly is to clamp it to a brass tube which slides up and down the support rod and is held in position over a clamp (Fig. 27). Both the clamp and the lower end of the tube should be serrated using a triangular file to prevent turning to and fro. This arrangement can be used wherever a complete assembly, such as a column, distilling head and condenser combined for solvent distillation, has to be movable in a vertical direction.

Filter Pumps (Aspirators)

This is perhaps the right place for some observations on this often troublesome topic. Ordering them is a hit-or-miss affair.

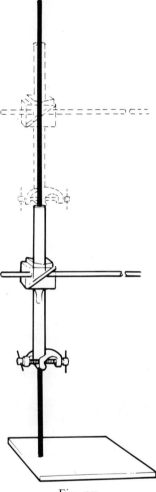

Fig. 27

You should (a) order only from a supplier who will take back and credit for unsuitable ones, and (b) order twice as many as needed on the assumption that half of them will turn out to be useless; this figure is based on bitter personal experience. A good filter pump, when connected to a manometer and a one-liter flask, should evacuate the system down to maximum attainable vacuum in less than 3 min. Metal pumps tend to be better than glass ones but have a shorter working life because of corrosion. All this is of course assuming adequate water pressure uninterrupted by sanitary operations and the like. In every case it is best to guard against unpleasant surprises by incorporating a guard vessel into the system; safety valves inside filter pumps reduce efficiency and often fail at the critical moment.

ON THE IMPORTANCE OF AZEOTROPES

Many organic reactions are equilibrium reactions in which the yield of the desired product can be enhanced by removing one of the constituents (water, an alcohol or another low-boiling product) from the system, and this is best achieved by use of an inert reaction solvent which can do so as an azeotrope. This principle is likewise important when freeing a product from traces of a high-boiling contaminant, and when recrystallizing from a mixture of solvents. A reference book such as the A.C.S. monograph on the subject[44] should be available to every research group; for the sake of convenience the accompanying Table 4 lists azeotropes formed by inert solvents with the most commonly encountered polar solvents or reactants. The figures in brackets in each entry show the approximate percentage by weight of the latter in each azeotropic mixture.

From this it is easy to see, for example, that the only way to remove traces of pyridine is with toluene, that traces of formic acid are removed much more easily than of acetic acid, and that ethyl acetate–cyclohexane or ethyl acetate–carbon tetrachloride are suitable solvent pairs for thin-layer chromatography and for fractional crystallization.

ON CHROMATOGRAPHY—COLUMN, TLC AND PREPARATIVE TLC

This wide field is covered extensively, in books, reviews, periodicals and in manufacturers' brochures. In gas chromatography and high-pressure liquid chromatography the over-

TABLE 4

Some important azeotropes

	Hexane 69[a]	Heptane 98.4	Benzene 80.1	Cyclo-hexane 81.4	Methyl-cyclo-hexane 100.3	Toluene 110.6	Carbon tetra-chloride 76.8
Methanol 64.65	50 (30)[b]	59.1 (51.5)	57.5 (39.1)	54 (38)	59.3 (54)	63.8 (69)	55.7 (21.6)
Ethanol 78.5	58.7 (21)	70.9 (49)	68.2 (32.4)	64.8 (31.3)	72.1 (47)	76.7 (68)	65 (16)
n-Propanol 97.2	65.6 (4)	87.5 (36)	77.1 (17)	74.3 (20)	86.3 (35)	92.6 (50)	73.4 (18)
Isopropanol 82.3	62.7 (23)	76.4 (50)	71.9 (33)	69.4 (32)	77.6 (53)	80.6 (69)	68.6 (28)
n-Butanol 117.7	—	94 (18)	—	79.8 (9)	95 (20)	105.5 (32)	76.55 (2.5)
s-Butanol 99.5	67.2 (8)	88.1 (37)	78.5 (15)	76 (18)	89.9 (41)	95.3 (55)	74 (8)
t-Butanol 82.8	63.7 (22)	78 (62)	74 (36.6)	71.3 (37)	78.8 (66)	—	71.1 (17)
t-Amyl alcohol 101.8	—	—	80 (15)	78.5 (16)	92 (40)	100.5 (56)	—
Acetone 56.5	50 (59)	56 (89.5)	—	53 (67)	—	—	56 (88.5)
2-Butanone 79.6	64 (29)	77 (73)	78.2 (45)	71.0 (52.5)	77.7 (80)	—	73.8 (29)
Tetrahydro-furan 65.5	63 (53)	—	—	—	—	—	—
Dioxan 101.5	—	92 (44)	—	79.5 (25)	93.7 (45)	—	—
Pyridine 115.3	—	—	—	—	—	110.1 (20)	—
Methyl acetate 57.1	51.7 (60)	57 (96)	—	55 (83)	—	—	—
Ethyl acetate 77.1	65 (38)	—	—	72.8 (54)	—	—	74.8 (43)
Dimethyl carbonate 90.5	67 (20)	82.3 (61)	—	—	—	—	75.75 (12)
Acetic acid 118.1	68.2 (6)	95 (17)	79.6 (2)	79.7 (2)	96.3 (31)	104 (32)	76 (1.5)
Formic acid 100.7	60.6 (28)	78.2 (56.5)	71 (31)	70.7 (30)	80.2 (46.5)	86 (50)	66.65 (8.5)
Water 100.0	61.6 (5.6)	79.2 (12.9)	69.2 (8.83)	69.56 (8.4)	81	84.1 (13.5)	66 (4)

[a] Figures not in parentheses refer to boiling point in degrees Celsius.
[b] Figures in parentheses refer to approximate percentage by weight of polar solvent in each azeotrope.

riding problems will be concerned with instrumentation and will thus be outside the intended scope of this book. The other field of primary concern to the synthetic organic chemist is that of adsorption chromatography (column and thin-layer). Here too there is no dearth of literature and much information on mainly theoretical aspects, but as far as information on practical application is concerned one cannot but feel that the greater part of the literature was not written from the point of view of the synthetic chemist or for his benefit. How else is one to explain, for example, the customary division into topics such as 'essential oils', 'alkaloids', 'lipids', or 'medicinal substances' (? !), instead of logical classification under functional groups?

Column chromatography

Columns and adsorbents

As is well known, for a given amount of adsorbent columns of smaller diameter effect better separation but have lower capacity and are slower to develop; as the diameter of the column is increased these properties are reversed. A compromise solution that has received comparatively little attention is that of the multibore column[45] as shown in Fig. 28. As far as I know, none of these are available commercially but their construction by a glassblower should not be much of a problem. In these the diameter of each portion is in the ratio of $c.$ 1.4 : 1.0 to the adjoining one. These columns are used extensively at some institutions, and in my experience they are invariably superior to the ordinary straight type for the same amount of substrate and adsorbent.

When working in a hot climate the use of a water-jacketed column (Fig. 29) may become imperative and may be good policy in general because it will cut down problems caused by exothermic adsorption of a substrate, such as local and sudden change in adsorbent activity and resulting column discontinuity.

There is probably no practical alternative to Teflon® stopcocks for chromatography columns, certainly for your peace of mind when a separation has to be interrupted, even for going to

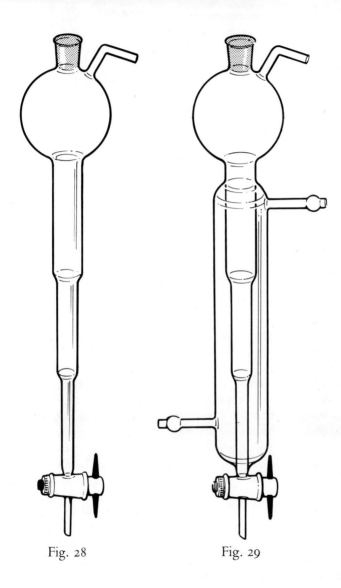

Fig. 28   Fig. 29

lunch. Glass stopcocks cannot be greased for obvious reasons and thus leakage can never be excluded. This also applies to spherical joints in which flow is interrupted by inclining the socket portion.

All adsorbents as sold by reputable manufacturers should be of activity I. A quick way to test for this[46] is the yellow colour shown on shaking a small amount with a 1% solution of chlorotriphenylmethane in benzene which is absent when the adsorbent contains 0.5% or more of water. For chromatography they have to be deactivated by mixing with the specified amount of water. Usually the advice given on the actual procedure is glibly stated as: 'add the required amount and then shake and wait for some time'. This is all very well if you or your assistant have plenty of time—and stamina. The best way is to do it in a closed ball mill jar (without the balls), or in a large flask attached firmly and in a horizontal position to a rotary evaporator (without applying vacuum). In both cases a speed of not more than one revolution per second should be used and two hours is usually sufficient. A fluorescent indicator can be incorporated at the same time, and for adsorbents impregnated with silver nitrate the latter can be dissolved in the water used for deactivation.

The way to fill a column depends on the adsorbent used. In the case of silica and Florisil a slurry in the initial elution solvent used should be stirred to expel air bubbles and until heat evolution has subsided; the amount of solvent used should be sufficient to enable pouring most if not all the suspension into the column in one go. For alumina it is best to add this in a fine stream, from a separating funnel with a non-greased stopcock, into the column containing a head of solvent of at least 3 ml per gram of adsorbent used. In all cases the column should be perfectly vertical and should be vibrated by hand or with an electrical vibrator until the adsorbent has completely settled.

The most critical part of a column is the very top portion. Its homogeneity and levelness predetermine whether the zones will elute in sequence or admixed with each other. For this reason it is inadvisable to disturb this layer by any means such as adding an additional layer of sand. The substrate solution should be added carefully by capillary pipette down the sides while there is still a sufficient solvent head above the adsorbent. Any resinous material which might clog the top layer should have been removed by a preliminary purification.

Elution order and choice of adsorbent

Roughly and in general, functional groups in an organic compound contribute to its retention, on the most commonly used adsorbents, in the following ascending order:

halide < hydrocarbon < olefin < ether < nitro < ketone < carboxylic ester < aldehyde < amine < tertiary amide < secondary amide or lactone < alcohol or primary amide < phenol < carboxylic acid.

On silica, amines, depending on their basicity, may be more strongly retained than is indicated. Another general factor is the relative proportion of the functional and the non-functional part of the molecule. Thus, a butyl ester is less strongly adsorbed than the corresponding methyl ester, and a steroidal hydroxy-ketone may be more easily eluted than a hydroxy decalone when the same adsorbent and elution solvent are used.

An important consideration when choosing an adsorbent is the stability of the substrate. Obviously base-sensitive compounds should not be chromatographed on basic alumina, and silica should be avoided in the case of very acid-sensitive compounds. The latter adsorbent is often used by less experienced workers to the exclusion of all others because they think they can 'see' the development of the column. This is usually a fallacy; discernible zones result in most cases from the refractive index of the solvent and not from the substrate. Moreover, the use of silica from any but the most reputable sources has been known, more than with any other adsorbent, to lead to rearrangements and other undesirable changes. In this respect Florisil is probably the most harmless, and magnesium silicate as supplied by Woelm or Merck, while inordinately expensive, can be better than most other adsorbents for separating mixtures of compounds differing but slightly in polarity.

As already mentioned, the great majority of books on the subject give practical guidance on the basis of case studies of groups of compounds of mainly commercial and pharmaceutical interest. This is not generally helpful to the synthetic organic chemist unless he can find an analogy to the type of compound he is working with, and this is rare.

The best policy is to file chromatographic information from the experimental part of the current literature. The most valuable source is usually from publications by large industrial laboratories where the pool of practical experience is likely to be greater than in university departments, particularly where intensive work on a particular class of compounds is going on.

A useful thing to keep in mind in the case of polyfunctional compounds is that separation can often be improved by modifying one or more of the functions. Thus, where separation of a mixture of epimeric nitro-ketones may be difficult it will usually be much better in the case of the corresponding nitro-ketals; and where a mixture of olefinic alcohols cannot be separated chromatography of the corresponding acetate esters on silver nitrate-impregnated silica will often solve the problem. In many cases such functional modification may well be part of the intended synthetic sequence (for example: ketone→thioketal→ desulfurization) and may even be planned with this separation factor in mind. As a general principle, the functional group which least contributes to the difference between the constituents of the mixture should be changed into a less retentive function. For example, the separation of a mixture of axial and equatorial hydroxy-ketones will be better served by conversion into the corresponding mixture of hydroxy-ketals than of the acetoxy-ketones.

Solvents for elution

Many books show a table of elutropic order of solvents, with some differences apparent from one book to another. *The Chemist's Companion*[47] gives a detailed list for alumina, silica and Florisil separately. It must be remembered that this information holds only for absolutely pure solvents. Traces of ethanol in both diethyl ether and chloroform may strongly modify the places of these solvents in the elutropic series.

A frequently given piece of advice is to determine the nature of the initial solvent or solvent combination by prior thin-layer chromatography. This may be good in principle, but in actual practice the best solvent combination for thin layers is not necessarily the most economical or convenient one for column separation. As a general principle one should tend to use mixtures of two solvents which differ widely in boiling point (to make recovery easier) and differ fairly widely in polarity, with the less polar one preferably of higher boiling point. Suitable such pairs are dichloromethane–hexane, ether–hexane and ether–benzene. The use of acetone should be avoided since this solvent may condense either with itself or with the substrate particularly when alumina is used.

Efficiency of separation and avoidance of 'tailing' go hand in hand with changing eluant composition as gradually as possible. Several books on the subject[48, 49] described devices for doing this linearly, i.e. 'gradient elution'; and wherever chromatography is done on an extensive scale it pays to use one of these.

For accurate and reproducible work it is necessary to use solvents which are 'isotonic' in regard to the absorbent, i.e. which contain a requisite amount of water.[50] Otherwise desorption of water from the adsorbent may occur on arriving at certain eluant compositions and lead to a change in its activity; this can be quite sudden and drastic.

Fraction collecting

There was a time when we witnessed a veritable feast of gadgeteering in this connection. This has now subsided to some extent perhaps because few such gadgets enable collection in flasks instead of test-tubes, or because all such gadgets are blind unless a still more expensive and not generally applicable method of testing fractions (UV absorption, refractive index) is incorporated. Such pieces of apparatus do pay off if you are continually working with the same type of compound and standard adsorbents and eluants and if you are certain of the constancy of your local electricity supply. Otherwise you should stick to the time-proven method of collecting and concentrating fractions by hand and by personal judgment, and analyzing them by thin-layer chromatography.

Fractions are best collected in long-necked round-bottomed flasks of 50, 100 and 250 ml capacity. Those without a standard joint cost only one-half to one-third of the standard joint variety. They should not be filled more than one-third full, and solvent removal with the rotary evaporator can be done using an adapter as illustrated in Fig. 30, which attaches to the flask by means of a piece of thick latex rubber tubing (probably best as far as resistance to solvent attack is concerned). They should be wired as shown with a ring, for suspension from hooks on a horizontal wire stretched along the bench shelf or framework, rather than taking up valuable space on the bench and using up a large number of beakers or cork ring supports. A TLC sample can be taken by lowering the capillary in its applicator with a long thin pair of forceps (if no sufficiently long ones can be obtained from a laboratory supplier try a medical supply store; they are useful also for recovering glassware from a chromic acid bath).

Thin-layer Chromatography

How did we ever manage without it?

On the question of whether to make your own plates or to invest in the commercially available ones it is hard to give any definite advice. The greater economy of the former may well be offset by their lower uniformity and hence unreliability and failure rate.

The commercially available ones are of glass or with a plastic or aluminum foil base. They are usually larger than the microscope slide in routine use and must be cut to that size or near it. With plates on a flexible base this is best done by a guillotine, with the adsorbent side facing down on a sheet of soft paper. Cutting with scissors often results in crumbling at the cut edge leading

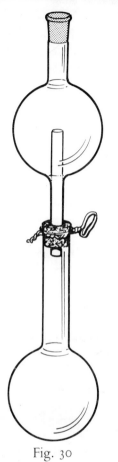

Fig. 30

to uneven ascent of the solvent front. Cutting glass plates is more of a problem. A plate-glass cutter which will do the job cleanly and evenly can only be found by trial and error. For this reason the prescored plates sold by Analtec Inc. are recommended. These measure 20 cm by 10 cm and from them sections corresponding to the widths of one, two, three or more microscope slides can be snapped off smoothly; these can accommodate up to three, eight and twelve spots respectively.

Alumina plates are usually more sensitive than silica ones to crumbling and mechanical damage and their cutting and storage demands extra care.

Spotting

This should always be done as uniformly as possible and reliance on home-made capillaries is not recommended. Graduated capillary pipettes are difficult to clean and dry completely and on balance it may be best to use the disposable type (e.g. the 'Micro-caps' sold by the Drummond Scientific Co. which come in vials of 100 together with an applicator). The most useful sizes are those delivering 1, 2 and 5 lambda; for extra economy the last-sized ones can be cut in half with scissors.

The use of a spotting template is strongly recommended. This can easily be made of glass or a plastic transparent material as illustrated in Fig. 31. It should be notched at one side at regular intervals and can include a lengthwise slot flanked by holes for application to preparative plates.

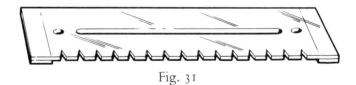

Fig. 31

Development

A point often made is that the use of a narrow rectangular vessel gives best results. However, most of the commercially available ones are for preparative plates and too large for routine work; also they are made of pressed or moulded glass and thus not sufficiently transparent. Thus one has to fall back on the use either of tall narrow spoutless beakers (the Berzelius type), or of suitable jam, jelly, marmalade or preserve jars to be found in any household. These must have as flat a bottom and as wide an opening as possible. Perhaps next time at the supermarket you should gently steer whoever is in charge of the shopping towards those brands which will eventually meet the requirements; and don't forget to take a tape measure with you. Naturally any vessel used must have a uniform rim for covering with a watch glass or Petri dish.

The development vessel must be saturated with solvent vapour at all times; this can be achieved by lining it with a strip of filter paper and shaking it frequently—not of course while a plate is being developed.

The best elution solvent combination is a pair mutually compatible, differing appreciably in polarity but little in vapour pressure. Suitable combinations can be chosen from solvents such as cyclohexane, benzene, ethyl acetate, carbon tetrachloride, acetone, chloroform and hexane. Bottles containing such pairs in different proportions should be kept ready for use. Frequently a pair containing diethyl ether may give better resolution but these cannot be kept for long (enrichment of the higher boiling partner, peroxide formation) and should be used only in a tightly closed vessel.

Thin-layer chromatography of carboxylic acids and enolic compounds usually gives rise to streaks and otherwise poorly resolved spots. In such cases you should try a mixture of a non-polar (e.g. ethereal) solvent and of formic or acetic acid. For example, mixtures of diisopropyl ether and acetic acid in proportions ranging from 99 : 1 to 90 : 10, on silica plates give particularly well-defined spots of carboxylic acids of higher molecular weight. Such mixtures must be kept in tightly closed bottles under argon in view of the special tendency of this solvent to undergo peroxidation.

Multiple development

This usually leads to greatly improved resolution, and in principle should be done as follows. Find out which solvent combination gives an R_F value of 0.4–0.5 after one development. Then use a combination in which the relative proportion of the non-polar partner is doubled and develop the plate in this three or four times, allowing the plate to dry lightly but not completely after each development.

R_F values

There is no sense recording these unless with reference to a standard substance run on the same plate and at the same time. Plate activity is very much dependent on temperature and humidity and hence on the weather and the seasons. Unless you are working in a laboratory which has automatic air conditioning these are entirely beyond your control.

Visualization

This is least of a problem when you happen to be working with compounds containing chromophoric groups. When these are run on plates whose adsorbent contains a fluorescent indicator, examination under an ultraviolet lamp (usually with 250 nm light) shows up the components as dark spots on a light orange background, a great advantage when having to decide if and how further development is necessary. However, unless all the components in a mixture are equally 'chromophoric' the picture can be very misleading. To be on the safe side such plates should always be compared with others which have undergone chemical visualization.

On the whole the iodine chamber is the most reliable standby even if the spots are impermanent. For this the plate must be oven-dried and then allowed to cool to room temperature, otherwise the spots may take a long time to appear. This method is applicable to almost all compounds containing some degree of unsaturation down to an isolated carbonyl group. It will fail with saturated hydrocarbons, and frequently with saturated esters, nitriles and epoxides. In such cases there is no alternative to spraying with a suitable reagent (e.g. acidic ceric sulfate), a list of which can be found in any book on the subject.

An interesting phenomenon is the negative spot (white on the brownish background) shown with iodine in the case of many halogen compounds.

Following a reaction by thin-layer chromatography

This is one of the most useful applications of the technique; and it is probably the mildest and most convenient tool available when trying to determine optimum reaction conditions (solvent, temperature, time, relative proportions of reactants) in a number of trial experiments.

Gas chromatography does have the advantage of giving a quantitative picture (on the rather doubtful assumption that on direct injection of a reaction mixture the course of the reaction is 'frozen'), but it takes far less time to find a suitable TLC elution solvent and plate than to establish the best type of GLC column, oven temperature and flow speed. Moreover, when using spectroscopic methods (IR, NMR) the reaction solvent and other components in the mixture will interfere and working up may be necessary.

There are indeed many instances where the conditions of a reaction whose course has to be carefully controlled should be planned with TLC follow-up specifically in mind. A high-boiling solvent will usually interfere in that it can never be dried off completely on the plate, and will then drag spots of components along with it and create a dreadful smear. Hexamethyl-phosphoric triamide, dimethyl formamide and dimethyl sulfoxide are particularly nasty in this respect. Sometimes this can be corrected by changing the nature of the plate and adsorbent. In fact, thought must always be devoted to the type of plate used. For example, when following the course of an acid-catalyzed reaction only alumina plates which instantly neutralize the catalyst give a true picture; on silica the reaction may proceed further on the plate itself and this at a rate significantly faster than in the reaction flask. Other cases where one has to think ahead are where one or more components are present not in the free state but as, for example, enolates. Sometimes instant protonation occurs on silica plates which solves the problem; in others it may be necessary to submit a drop of the mixture to working-up in a micro-centrifuge tube.

Always keep on hand TLC samples from every significant experimental run: starting material, intermediate stages and crude product; until your project has come to a conclusion, either in the form of writing up for publication or by being irrevocably abandoned. You can be sure that there will come a time when you will wish you hadn't thrown that sample down the sink. Obviously this is what to do with leftover or recovered sample solutions from NMR or IR examination. They are best kept for that purpose in screw-capped one-dram vials—these take up very little space and are still large enough to carry a legible label or serial number. For longer periods they must, of course, be kept in the refrigerator.

On preparative thin-layer chromatography

This technique is now routinely used for quantities of up to 300 mg and is probably the fastest and most efficient method if you are lucky to work with compounds whose visualization is no problem. When the components are coloured in the visible region one is often tempted to preserve the plate for posterity as the *dernier cri* in post-impressionist painting (anyone wanting to go in for this seriously is advised to use crude products from an oxidation with dichlorodicyanoquinone). The difficulty arises with compounds having no chromophoric groups. Here there is probably no alternative to dividing up the plate systematically into sections which are eluted separately. Subjecting a small section of the plate to chemical visualization is acceptable only on the assumption, generally fallacious, that all the zones are parallel. This happens only with plates of the highest quality and where the substrate solution is applied with an automatic device. A closer approximation can often be made by applying the mixture in sections separated by single spots (at each side and in the middle) whose location is carefully marked and which are then subjected to spray visualization after covering up the main portions.

An important thing to remember is that adsorbed compounds are very sensitive to aerial oxidation, and more so at higher temperature. Hence plates should preferably be dried as quickly as possible after development in a vacuum desiccator if indeed they have to be dried at all. This is particularly important when they are subjected to multiple development.

Elution of scraped-off sections of a preparative TLC plate is best done by Soxhlet extraction. The thimble section should be of as small a volume as possible to cut down on the amount of solvent used; if necessary dead space can be filled up by adding clean glass beads. Many extraction solvents may dissolve or otherwise carry over traces of the adsorbent, and hence the solution should be concentrated, a less polar solvent added, and the whole filtered through a very short column inside a pipette.

Finally: a preparative plate takes anything between one and two hours for one development. It follows that when several developments are needed (and one should, if at all possible, finish the same day), one has to (a) start at 7 a.m. (which is always the best time in the lab) and (b) plan to do something else constructive in the vicinity while development is taking place.

ON GETTING YOUR PRODUCT CRYSTALLINE

This is where the men are separated from the boys and where organic chemistry is not a science at all. It was said of Adolf von Baeyer that his success was in large measure due to his large beard harbouring seeds of every compound ever made. However, even if beards are in fashion again this alone is not a good enough reason for following his example.

Some people believe in mascots or singing operatic arias; at least it cannot do any harm.

In giving any kind of concrete advice on the subject one can only cite one's personal experience; and mine is that the use of diethyl ether as solvent over the entire temperature range from $-80\ ^\circ C$ to its boiling point, together with scratching with a micro-spatula, is still the best thing to try first. Another 'method' worth attempting is to cover the material with a solvent in which it seems insoluble (e.g. hexane), cool to $-10\ ^\circ C$, push a blob of it up the side of the flask, partially dissolve this with a drop of ether running down the side, and scratch immediately.

If nothing else seems to work and you are beginning to lose your temper, leave your product over the weekend covered with a solvent in which it seems to be partly soluble, and, shocking as it sounds, leave the flask open. Slow evaporation together with nucleation by dust frequently does the trick. This is better than leaving it in the refrigerator; higher viscosity at low temperatures is not helpful in inducing crystallization. Obviously, crystallization should be seriously attempted only if the product appears to be reasonably homogeneous by TLC or GLC, and preferably after preliminary purification (Kugelrohr distillation etc.).

On Recrystallization

When finally collecting your product for recrystallization, whether after chromatography or other method of purification, the size of the flask should bear a reasonable relation to the total amount, for instance 50 mg should never be collected in a flask of over 25 ml capacity. While doing this, never forget to keep a seed crystal (from one of the chromatographic fractions). The first crystallization should then be made in the flask itself. Place it with the minimum amount of solvent on the smallest steam bath opening (if no small ring is available make one yourself) and allow the rising and condensing solvent to wash down material adhering to the side. The flask should then be upright while cooling; for this you should always be on the lookout for suitable circular objects (rings from spent adhesive tapes) since beakers are rarely of the right size for this. After cooling, mother liquor removal and washing should be by capillary pipette only.

Subsequent recrystallization should always be in a small Erlenmeyer flask—never in a test-tube or beaker. The Craig tubes still used by some workers appear to have gone out of fashion, probably for good reason. It always pays to choose a solvent or solvent combination from which the product separates in compact form allowing for nearly quantitative mother liquor removal by pipette. And even if you do get a mass of fluffy needles use a pipette with a thin stem and withdraw solvent right from the bottom; the crystal mass will itself act as filter. If filtration of the solution is necessary use only filter paper and never cotton wool (this always contains some soluble material) nor sintered glass (you have no way of knowing what is left inside the pores). As for funnels, the best are those with a wide stem which allow boiling solvent to dissolve and wash down material crystallized on the paper and funnel rim, and with an angle of 45° instead of the 'regulation' 60° (Fig. 32): none of these are commercially available in micro-size (20–35 mm upper diameter), and an adequate supply should be made by a glassblower.

Drying is best done in a vacuum desiccator, but not before the flask opening has been covered by one or two layers of aluminum foil punctured in several places and held by a rubber band. Otherwise the crystals may jump right out due to sudden evaporation of adhering solvent, and when air is reintroduced.

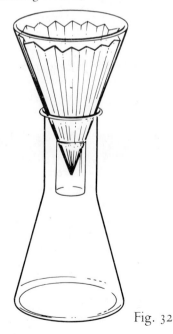

Fig. 32

Low Temperature Recrystallization

There is no reason to shun this in small-scale work once you have mastered the use of the capillary pipette. The flask containing the crystallizing solution, and another flask containing solvent for washing and closed with cotton wool, are placed in the deep-freeze compartment. After crystallization is complete the mother liquor is quickly withdrawn; the flasks are then cooled in a carbon tetrachloride–dry ice bath and with another pipette the washing solvent is rinsed in down the sides. After swirling it is again withdrawn and the process is repeated if necessary.

Even with larger amounts low temperature crystallization using capillary pipettes is preferable to collecting crystals by rapid filtration, because this is where losses are always incurred. Any adhering solvent should be quickly removed at the rotary evaporator. It is true that remaining liquor which cannot be removed completely in this fashion may lead to a product which is less pure than it might be after filtration and washing on the filter, but the process can always be repeated and the mechanical loss is nil.

Every low temperature crystallization should preferably be done in stages, e.g. first in the main refrigerator compartment ($+4$ °C) for some time and only then in the freezer (-35 °C).

Fractional Crystallization (Triangulation)

When the organic chemist of today is faced with having to separate a mixture of crystalline products differing but little in polarity, his first impulse is to turn to yet another instrumental separation method or a refinement of it: another type of column in gas chromatography, high-pressure liquid chromatography and so forth. As a result the art of systematic fractional crystal-lization is dying out, or at least it is mentioned less and less; and this is rather a pity not only considering the sense of aesthetic satisfaction it can provide but also because for larger amounts it is still the most practical way. To refresh your memory Fig. 33 shows the classical scheme in more intelligible form.

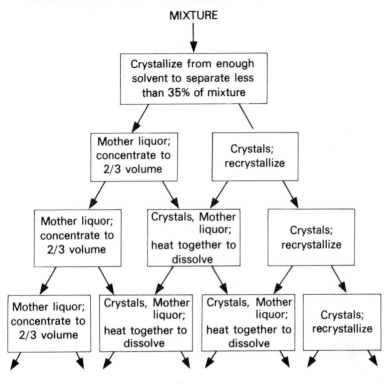

Fig. 33

Naturally the sequence is interrupted at any stage where melting point determination or spectroscopic examination show that a pure product has been separated. Any 'unpartnered' mother liquors left as a result should then be combined to start the process afresh, with 'seeding' where judged appropriate.

A number of reservations have to be made:

(a) Cleanliness is of the highest importance. Hence solvents must be absolutely pure, and the initial mixture must have undergone preliminary purification. Mother liquor withdrawal and washing of crystals should be by pipette only, and the use of boiling stones should be avoided.

(b) Only a single solvent should be employed, or possibly a pair of solvents of similar vapour pressure. The solvent should be stable to air, which leaves out ethers, and non-hygroscopic, which eliminates most alcohols. Thus the choice is rather narrowed down.

(c) Careful judgment has to be employed in determining when to proceed to the next stage, i.e. when to stop the crystallizing process and transfer the mother liquor, and this comes only with experience. Crystal shape (examination with a magnifying glass) is an important criterion.

(d) While crystallization proceeds the vessel must be left absolutely motionless and the usual temptation to pick it up for examination has to be resisted. It is best to place it on a thick glass plate with a dark backing—this may have drawn on it a diagram on the lines of Fig. 33.

ON USING THE KUGELROHR APPARATUS

This exceedingly useful equipment should by now be in standard use even in undergraduate laboratories. It can be employed for distillation and sublimation of amounts from 5 mg to over 30 g (see below). It is the fastest and most convenient way to purify preliminarily even the most messy-looking crude product.

The tubes used (Fig. 34) can be made by any reasonably skilled glassblower. The distance between bulbs should preferably be sufficient to permit the use of a rotary glass cutter to disconnect

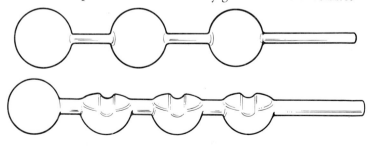

Fig. 34

them. They are filled by placing them vertically on a steam bath and adding the material dissolved in ether or in dichloromethane by capillary pipette from the top, followed by washing down with more solvent. Most of the latter should evaporate off instantaneously (use a boiling stone) and the rest is removed by gradual application of vacuum while gently swirling the tube.

Some workers prefer bulbs connected by standard joints (Fig. 35), but it should be remembered that grease, especially when hot and subjected to vacuum, will flow inward and contaminate the product. The use of a device to rotate the tube during distillation is often of great advantage but this involves extra cost, is space-consuming and if extraneous to the apparatus, will reduce flexibility in the alignment of the tube. One manufacturer has now come out with a model in which the rotation mechanism is an integral part of the oven.

Fig. 35

When the amount to be distilled is near the maximum capacity of the bulb it is advisable to incline the oven and tube at first and then gradually return them to horizontal as the distillation proceeds. Cooling of the receiving bulb can be done in most cases by a swab of cotton wool soaked in methanol, and if necessary by pieces of dry ice held in place by a boat made of stiff aluminum foil. In the hands of skilled workers it is possible to achieve a degree of fractionation by distilling into the farthest bulb at the lowest temperature and then gradually exposing additional bulbs as the oven temperature is raised. Transfer of the bulb contents, e.g. for crystallization, is done as shown in Fig. 36, allowing the rising solvent to dissolve and wash down the material; excessive loss of low-boiling solvent can be avoided by holding a small piece of dry ice to the top part (arrow) of the bulb. After breaking off and removing the contents it pays to clean the bulbs and save them for reconnection.

Fig. 36

Three ways of exploiting the potentialities of the apparatus for larger quantities are illustrated in Figs 37, 38 and 39.

The dimensions are limited only by the internal ones of the Kugelrohr oven. In Fig. 37 the upward bend (which should stay upward!) between the distilling and receiving bulb serves to avoid contamination by splashing. Figure 38 is self-explanatory, while the simplest arrangement which can be made from any test-tube (Fig. 39) calls for some comment. When the product,

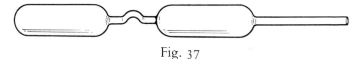

Fig. 37

collected between the indentations, is crystalline, it can usually be scraped out using a spatula with a sharpened edge or a scalpel, but when as in most instances it is a viscous gum there arises the sixty-dollar question of how to transfer it without contamination by the residual tar. I am indebted to Professor G. Stork (Columbia University) for an ingeniously trivial solution to this problem: carefully fill the tube held upright with clean mercury up to the

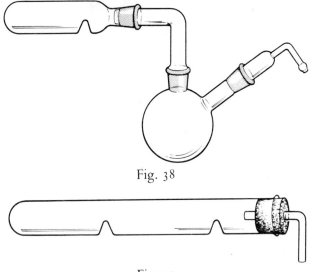

Fig. 38

Fig. 39

bottom indentation, add solvent on top, allow the product to dissolve and withdraw the solution by pipette. But don't forget to clean the mercury before using it again!

ON FRACTIONAL DISTILLATION

When one considers the glassware encountered in most laboratories and offered by the majority of glassware manufacturers one is struck by the lack of suitable apparatus for fractional distillation of quantities of between 1 and 15 ml. This range is, of course, between that commonly dealt with by preparative gas chromatography and the quantities encountered in the teaching laboratory.

For very accurate work there are spinning band columns, and micro and semi-micro versions are commercially available. However, they are very expensive and their repair is a task which even experienced glassblowers are reluctant to undertake.

Most advanced workers solve this problem on an *ad hoc* basis, by asking the glassblower to make up an apparatus for a particular distillation on hand, after which it is often relegated to a far corner of the drawer and its existence kept a dark secret.

The main problem in fractionation on this scale lies in loss of material, caused by (a) insufficient immersion of the stillpot, (b) too long a distillation path and flooding, caused by too many bends and joints and constriction by the latter, (c) excessive hold-up due to inappropriate column filling, and (d) further loss on the way caused by impeded outflow, unnecessary use of a condenser or use of one with excessive internal surface. To meet these objections the use of the apparatus illustrated in Figs 40 and 40a is suggested. The distillation flask, preferably of the shape shown earlier on in Fig. 3, contains a small stirring bar and can be of 5–50 ml capacity. The material to be distilled is introduced through a tall (to enable maximum immersion) side tube of diameter suitable for closure by the smallest rubber septum available. The height of the column, and its diameter and that of the side arm, should range from 70 mm and 7 mm respectively for a 5 ml size to 150 mm and 13 mm for the 50 ml version. Length (p) from exit to the standard 14/23 joint should be the minimum needed for cooling by a methanol-soaked swab of cotton wool. With the slow rate employed in distilling on this scale 40–45 mm is sufficient to cool liquids of b.p. 45 °C and over with little loss. Dimensions (b) and (c), from the end of the joint to the drip tip, are the same for all sizes, and are designed to fit a standard fraction collector (see below).

Probably the best column filling for fractionation on this scale is that of the twisted wire gauze type described by Bower and Cooke[51] who give simple directions for making these from a strip of monel or stainless steel wire gauze and an identical pair of flat-nosed pliers. This type of filling leads to separation which in all cases is better than with any Vigreux column of comparable size, with minimal hold-up and pressure gradient; its performance is described in some detail by Krell.[52] It is slowly attacked by acyl and allylic halides but, unlike the vastly more expensive spinning band columns which suffer from the same problem,

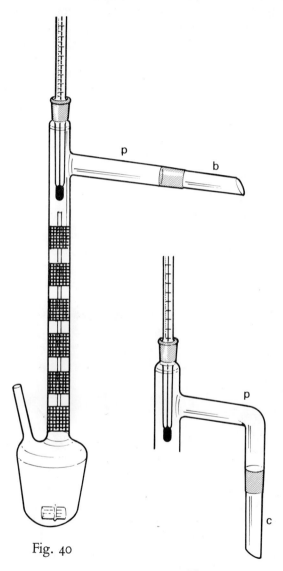

Fig. 40

Fig. 40a

it is much cheaper and made more easily. Naturally the column can be fitted with a vacuum jacket; silvering is not advised because it really adds little to performance while literally keeping you in the dark.

The fraction collector shown in Fig. 41 is designed for both versions of the above flask; in either case the vacant socket joint is used for vacuum take-off. As can be seen, it consists of two halves connected by a flange joint which must be carefully

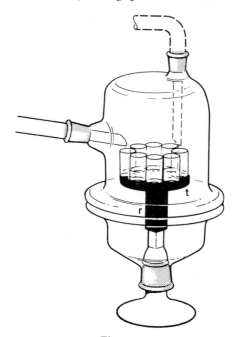

Fig. 41

greased and protected from heat. The lower half contains a ground-in key for turning, on whose upper square end rests a shallow metal tray (t) by means of a square hollow tube (r). The latter can be packed with spacers to allow variation in height. On the tray, vials of the rimless type and of different sizes can be arranged circularly and kept in place either by using the mainspring of a clock or by means of a suitably sized vial in the center. The tray can accommodate up to eight collection vials. The whole assembly is best supported by a ring support.

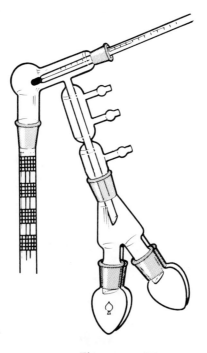

Fig. 42

Where efficient cooling is necessary, a set-up utilizing a micro-condenser (Ace Glass Cat. No. 9125 or Kontes Glass Co. Cat. Nos. K 284800, K 287100), as illustrated in Fig. 42, can be used, together with a 'cow' type fraction collector; incidentally this is called 'Spinne' (spider) in German-speaking countries. Its capacity is limited to the collection of four fractions unless made by a very experienced glassblower.

PREPARING SAMPLES FOR ANALYSIS

By Crystallization

All glass and porcelain ware used must be cleaned in chromic acid, washed with distilled water, dried in the oven in a clean covered beaker and stored in dust-free screw-capped jars.

A 10 ml Erlenmeyer flask with a 25 mm diameter micro-funnel and fluted filter paper (see Fig. 32) is rinsed with the pure solvent employed to remove traces of dust and placed on the steam bath. The hot solution is then added by capillary pipette through the filter, allowing the hot rising solvent to wash down material adhering to the paper. Such final recrystallization is best done using a solvent pair such as dichloromethane–hexane or benzene–methanol. No boiling stone should be used; if absolutely necessary the open end of a melting point capillary may be inserted to assist ebullition.

The crystalline material is collected on a 30 mm upper diameter Hirsch funnel, after which it is carefully transferred into a shell vial of 10 mm diameter together with the filter paper. The latter is then gently lifted out using a micro-spatula (never scratch the material off the paper!). The mouth of the vial is closed with aluminum foil punctured with a needle (not paper-clip) and dried in the drying pistol.

Analytical samples are generally sent by mail and often by air mail. The sample tubes often advocated have by weight and size an unreasonable relationship to the sample weight generally required. Hence, after drying, the requisite amount should be transferred by micro-spatula to a small (7 mm by 30 mm) vial which is then tightly closed by a cork stopper (made pliable with a cork press) covered by aluminum foil. The screw-cap half-dram vials often used have a lining which cannot be relied upon for cleanliness.

Liquids

A sample from a middle fraction, and/or after Kugelrohr distillation, is transferred by a clean and dry capillary pipette to the bottom of a clean dry Pyrex® test-tube (9 mm by 70 mm), taking great care not to wet the side. The tube is then quickly sealed using a small hot flame, positioning both tube and flame as shown in Fig. 43. If necessary, the tube can be evacuated and filled with inert gas, using the multi-outlet nitrogen trap, before sealing.

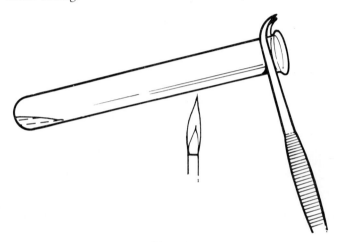

Fig. 43

Sublimable Solids and very Viscous Liquids

Here the procedure can be much simpler. A glass tube of 7 mm outer diameter is closed at one end to a small bulb which can be allowed to hang down, which is what happens anyway to a less skilled glassblower like yourself. This is placed vertically on the steam bath and the sample (10–30 mg) is introduced from the top by capillary pipette in solution as described for Kugelrohr tubes above; washing down with more solvent and evaporation of the latter must be done with particular care. The sample is then distilled or sublimed using the Kugelrohr oven in such a way (tube slightly off horizontal) that the material collects in a concentrated ring of crystals or drop of liquid. The tube is then sealed as shown in Fig. 44, first at point (A) and then at (B). Enough length of tubing should be left on either side of the sample to enable the analyst to break the tube without contaminating the sample with broken glass or with the adhesive of the label.

Fig. 44

SENDING SAMPLES TO FRIENDS AND COLLEAGUES

Many people are not yet aware of the advantage of having an instrument for heat-sealing polyethylene bags, not only for the above purpose but also for storing and mailing small quantities of solids without fear of breakage. Handy versions of such a device (e.g. see Fig. 45, overall length 35 cm) can be found in the household or electrical goods sections of larger department stores in Germany and Switzerland and in the U.S. (but not in the U.K.); prices at present range from $20 to $35.

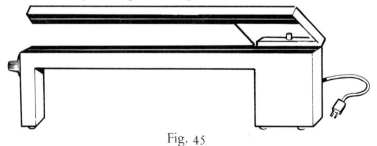

Fig. 45

Such bags should be sealed with the material and the label in separate compartments as shown in Fig. 46. The finished product weighs little more than the sample itself, and can be attached to the letter with adhesive tape (but not stapled!).

However, sending analytical samples in this manner is not advisable.

Of course with liquids there is probably no way to get around the use of a sealed ampoule.

Fig. 46

4
Some Basic Safety Rules

1. The moment you enter a laboratory where any work is being done, whether by yourself or by others, put on *Safety Glasses*. This should become a conditioned reflex. Perhaps the best way to induce it is a system of on-the-spot-fines which everyone is entitled to collect; the proceeds to go to some worthy cause like the Coffee Club.

2. At least once a day while in the laboratory stop and ask yourself what you would do should an accident or fire occur *at that moment*:

> Where are the nearest eye-flushing device and safety shower? Do they work?

> Where is the nearest fire extinguisher? When was it checked last, and by whom?

> Where is the nearest sand bucket? Does it contain sand or cigarette butts?

What emergency escape route is open to you from where you are working? You should refuse point-blank to work in any location from which there is only one way out.

Where is the nearest protective mask? The location of these should be clearly marked—the most important one is a smoke mask.

Are the gas cylinders secure? In an emergency it may be necessary to move them away as quickly as possible, hence it is better to have them secured on movable carts rather than strapped or chained to the bench.

Where is the main gas valve?

Which is the most direct route to a first-aid station?

What is the telephone number of the Fire Brigade?

3. *Medical treatment.* Few doctors have enough chemical or toxicological knowledge to be able to judge which is the best treatment for poisoning, injuries or burns caused by a specific chemical. Many charts are available which suggest the best course of action in common cases, and these should be available at all times and shown to the doctor or nurse. In most large hospitals there is now a Poison Information Center which can be reached at all hours, and its telephone number should be posted.

4. Do you always keep in mind that you can be held personally accountable for the consequences of allowing *children* and other *unauthorized persons* into the laboratory?

5. Avoid *working alone*, and never work in the absence of someone else on the same floor or wing who knows of your presence and whereabouts and is within shouting distance.

6. Always tend to work in a *hood*, especially where there is any element of risk. All organic solvents should be regarded as toxic on principle. Relative toxicity values should not be relied upon—you have no way of checking how much of each you inhale or handle during any specified period.

7. *Overnight reactions* involving heating and/or reflux and/or stirring should be done in a Night Room. This should have a device cutting off electricity in case of interruption in the water supply. Plastic rather than rubber tubing should be used for water cooling (less risk of splitting or breakage). Wiring-on should be done with care. Leave a securely fastened note with your name, address and telephone number next to each experiment.

8. *Disposal.* It all depends on what is being disposed of, the state of the plumbing, the location of your institution, and above all the laws and regulations of your community and state. Whole books can and should be written on this subject. Here is just one cardinal point: whatever you want to throw away and whichever way you decide to do it can have far-reaching legal and other consequences *for you personally.*

It should be the duty of your supervisor and of the administration of your institution to formulate clear and unambiguous disposal instructions and to assume full responsibility.

9. Every laboratory should have posted *on the outside* a list giving names, addresses and telephone numbers of people to contact in case of fire, floods or similar occurrences.

10. *Refrigerators* and *Cold Rooms.* These may malfunction, and catch fire or explode, *at any time.* It follows that this is more likely to happen *outside 'normal working hours'.*

This statement is not based only on the laws of probability. Its truth will be confirmed by anyone with long enough experience, and should always be kept in mind and not least by those 'in charge' who tend to economize, e.g. by pensioning off the nightwatchman.

No open vessel should be placed inside. To be quite sure, light bulbs should be taken out and any other potential source of sparking should be securely insulated.

5
On Catalytic Hydrogenation

The chemical, and to some extent the practical, aspects of this subject are dealt with in a number of books among which should be mentioned those by Augustine,[53] Freifelder,[54] Rylander[55] and Zymalkowski.[56] The following section deals with some points that are not generally mentioned or elaborated upon.

CATALYSTS

The above-mentioned books as well as other sources such as *Organic Syntheses*[57] and *Reagents for Organic Synthesis*[58] contain much information on preparation of catalysts, based on metals of the platinum group and others and on a variety of supports. Many varieties are available commercially, the price being variable and depending on the state of the rare metals market. The problem is that many of these either are pyrophoric or else lose their activity after a short time. A bottle of catalyst usually stands around for a number of years before being used up, and will have been opened (and imperfectly closed) by a large number of people before that time. In view of this the range of platinum metal hydroxide catalysts whose preparation is described by Pearlman[59] is highly recommended. These are non-pyrophoric and stable more or less indefinitely. They must be pre-reduced immediately before use, which means that they are then always more active than a ready-made catalyst and that their activity can be relied upon. As support for these only charcoal, purified by treatment with dilute nitric acid, should be used; and as has been pointed out they retain water tenaciously; they have to be dried in a high vacuum for several hours at 60 °C (the outlet should be plugged with cotton wool to prevent loss due to spattering into the vacuum system).

CATALYST RECOVERY

This is a subject much like the weather; many people talk about it and religiously save up residues, but few ever seem to do anything about them. Now I must confess that I have not read each of the above-mentioned books on catalytic hydrogenation from cover to cover, but I have a sneaking suspicion that the subject of catalyst recovery is studiously avoided in all of them. Perhaps this is because their authors have long known, and kept to themselves, that the amount of credit which a supplier is willing to extend for returned residues will barely exceed the minimum he charges for assay and recovery costs unless truly industrial quantities are involved.

Nevertheless, you should save used catalysts and recover them yourself when you have nothing better to do; or alternatively induce one of your inorganic colleagues to include a recovery exercise as part of his course. It is unfortunate that for some reason (once again ! ?) there are few if any books that give detailed directions; but there are sources, such as Gmelin's *Handbook*,[60] *Inorganic Synthesis*[61] and Mellor's *Treatise*,[62] from which enough information can be deduced to enable you to devise a workable procedure. If ever you write a book about it you will earn the gratitude of many, even while possibly incurring the enmity of a few.

HYDROGENATION AT ATMOSPHERIC PRESSURE

The types of apparatus used are all more or less based on the same principle: one, two or more gas burettes connected by a manifold to a hydrogenation vessel, and to a source of hydrogen and of vacuum. The burettes should be filled with dilute copper sulfate solution to prevent the growth of algae and other forms of life. The use of mercury is not advisable; there will always be some part of the apparatus which will not stand up under the weight. Moreover, one can never rule out the possibility that some local idiot may try to mop up spilled mercury near a hydrogenation set-up with powdered sulfur.

Using ball and socket joints to connect the various parts of the apparatus makes for flexibility, but leakage always tends to affect one of them and is difficult to trace. No rubber parts should be used anywhere; there is no existing elastomer which is impervious to hydrogen. That should rule out the use of rubber septa, and leave only magnetic stirring as a means of agitation, except where connection of the hydrogenation vessel is by means of a glass spiral in which case shaking can be employed to a limited extent.

Flasks used as hydrogenation vessels are generally of the *ad hoc* variety which do not fulfil the basic requirements: using as few joints as possible and enabling rapid and yet steady magnetic stirring. For these reasons the flask shown in Fig. 47 is recommended. In this the stopcock must be of the highest quality. The joint (j) connecting to the hydrogenation system must be male to prevent grease contamination. The procedure to be used with this flask is as follows: the catalyst and stirring bar are barely covered by the solvent added from the funnel (f) leaving one drop above the stopcock, and the flask is connected

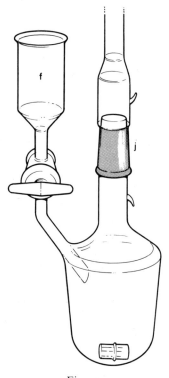

Fig. 47

to the system. The whole is then evacuated and filled with hydrogen three times, after which the catalyst is allowed to prehydrogenate until uptake ceases completely. A partial vacuum is then created in the system by lowering the burette reservoir, and the substrate solution is allowed to flow in from funnel (f) and is washed in with more solvent, taking great care not to allow air to enter the system. After this, hydrogenation is allowed to proceed at or slightly above atmospheric pressure.

Another advantage of this flask is that it can be immersed almost completely in a constant-temperature bath.

HYDROGENATION AT MEDIUM PRESSURE (UP TO 4 ATM.)

For this there is surely nothing better than the Parr hydro-genator,[63] and it is probably true to say that the advent of this apparatus and of rhodium and ruthenium catalysts have made most laboratory high-pressure hydrogenations a thing of the past. This instrument incorporates a tank containing hydrogen under pressure which is connected via a flexible polypropylene tube to a thick-walled glass hydrogenation bottle, via a manifold which also enables evacuation of the latter. The bottles come in 250, 500, 1000 and 2000 ml sizes via suitable spacers, they are shaken by a spark-proof motor and are enclosed by a wire-netting safety shield. Hydrogen uptake is shown by the pressure drop as indicated by a gage connected to the system, and can be calculated on the basis of the volume of the tank and the free volume inside the bottle, and the temperature. Some models allow for heating, though exact temperature measurement is a problem unless a bottle with a thermometer well can be used.

Later models have two pressure gages, one for the tank and the other connected to the bottle only; and perhaps it is not generally realized that this makes it possible to follow the hydro-genation of small amounts of substrate by closing the connection to the tank. Uptake, shown by the bottle gage, can then be calculated on the basis of:

$$\text{Pressure drop at N.T.} \atop \text{(p.s.i.)} = \frac{354}{\text{free volume}} \times {\text{No. of millimoles} \atop \text{of } H_2 \text{ absorbed}}$$

where free volume is the volume of bottle and connecting tube, less total volume of solution, in ml. For very accurate work uptake by catalyst and solvent must be determined by a blank experiment.

The most frequent source of leakage in this apparatus is where the flexible tube enters the glass bottle via a one-hole rubber stopper. It is quite futile to try to use home-made stoppers, and those supplied by the manufacturer should be used exclusively.

GENERAL

At the beginning of each hydrogenation air must be evacuated and then hydrogen admitted three times; and at the end all the hydrogen must be evacuated before air is readmitted. Failure to follow this practice punctiliously can have serious consequences.

Purity of solvents and in particular of the substrate are of critical importance. A melting point is not necessarily a criterion for the presence or absence of impurities which can act as catalyst poison. If at all possible material to be subject to hydrogenation should be distilled or sublimed as well as recrystallized. If it has been at any stage in contact with a sulfur- or selenium-containing reagent it should be stirred in solution with finely divided precipitated silver powder or, better still, passed through a column of a suitable adsorbent admixed with up to 10% silver powder. In chromatography this may conveniently constitute the lower part of the adsorbent column. A hydrogenation which because of poisoning appears to take days for completion either shows so-called uptake because of leakage or diffusion, or else will be incomplete. Far better to stop, filter, use fresh catalyst or repurify the substrate!

There is no need to have the substrate completely in solution at the outset when, as is frequently the case, the hydrogenation product is more soluble. Besides a hydrogenation reaction by its nature proceeds more slowly in dilute solution.

6
On Keeping it Clean

It should become an automatic habit to clean glassware after use preliminarily by dissolving out residues, before washing with a water-based detergent. This is the best use to which to put recovered solvent mixtures.

Drying glassware is preferably done in an oven if there is room for all. If not, it has to be allowed to drain. The customary pegboard has disadvantages which are obvious with many types of item. A much better installation for the purpose is a horizontal drying rack (Fig. 48) which incorporates a framework of thin plastic-covered rods forming square or rectangular openings of various sizes. Here there is no contact with the inside of the apparatus, and this rack is suitable not only for flasks but also for funnels and flat objects. If not commercially available it should be possible to make one in any workshop.

Although it is supposed to show up better in photographs when dirty, the necessity for using clean glassware, particularly

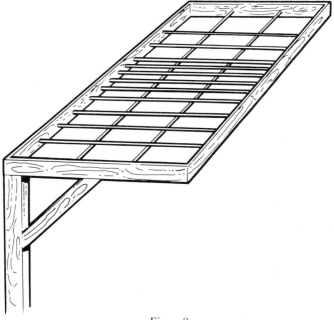

Fig. 48

in small-scale work, is generally not questioned. But when it comes to porcelain ware: 'What the eye does not see the heart does not grieve over.' One should make it a habit of dissolving and washing through material adhering to a Hirsch or Buchner funnel into the vessel holding the mother liquor. To make quite sure, all porcelain ware should be cleaned regularly in chromic acid.

And now to a more delicate observation. In many institutions the demand for cleaning tissue reaches alarming dimensions at which point it becomes clear that it is being used for purposes best described as non-scientific. I now suggest a reasonable and far more economic alternative: toilet paper. In principle it is exactly the same. You will soon get over the psychological block; and if the secretarial staff do not, so much the better.

7
Some Detailed Reaction Examples

The following, from unpublished work done in the author's laboratory, are chosen to exemplify some of the points dealt with in this book. The stereochemical assignments for products in examples 3 and 4 will (hopefully) be discussed in future publications.

I. Alkoxycarbonylation of a ketone, followed by *in situ* alkylation

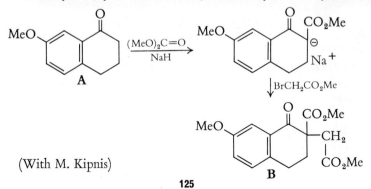

(With M. Kipnis)

Apparatus: Two-necked 50 ml flask with magnetic stirring bar, short column and small reflux-distillation head connected via silica gel drying tube to nitrogen trap (Fig. 14, p. 42) (arrow) (Fig. 49).

The apparatus was flamed out (from near the drying tube towards the flask side neck) under positive nitrogen pressure which was then maintained throughout. After cooling, sodium hydride suspension (assumed to be 57%, 343 mg, 8.15 mmol) was introduced and washed (p. 63) three times with hexane and then once with dimethyl carbonate (Note 1). A total of 20 ml of dimethyl carbonate was added, and then the ketone A[64] (dried over P_2O_5, m.p. 63–63.5 °C, 1.250 g, 7.09 mmol). The mixture was stirred and reaction initiated by adding one drop of absolute methanol and by warming (oil bath) to 65–70 °C. When hydrogen evolution had subsided the bath temperature was gradually raised and the methanol–dimethyl carbonate azeotrope (Note 2) was very slowly distilled off until the distillation temperature remained above 89 °C and 5 ml distillate had collected (1.5 h, Note 3). The apparatus was then raised from the oil bath, allowed to cool and methyl bromoacetate (redistilled, 0.83 ml, 1.36 g, 8.87 mmol) was added. The suspension was heated under reflux (S closed) with stirring for 3 h, cooled, the flask was disconnected, grease was removed, ice and ether were added; and after stirring for 5 min the aqueous layer was separated (pipette) and again extracted (Erlenmeyer flask) with 10 ml benzene. The organic layers were washed with N NaOH and with saturated NaCl solution, combined, dried (Na_2SO_4) and the solvents evaporated (rotary evaporator). A small portion of the residue was induced to crystallize by scratching with hexane at 0 °C and a seed crystal retained at the tip of a micro-spatula. The product was dissolved in dichloromethane (3 ml), hexane (5 ml) was added and the solution was passed through a column of Florisil® (3 g) which retained a dark impurity; the product was washed through with dichloromethane and then with chloroform until no more was eluted. After solvent removal (finally at 100 °C/0.1 mm) the residue was transferred in warm diisopropyl ether (pipette,

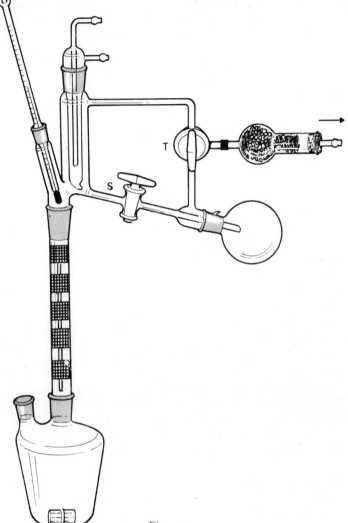

Fig. 49

Note 4) into a 25 ml Erlenmeyer flask and the solution concentrated to a volume of 10 ml. After cooling, it was seeded and the
product allowed to crystallize, finally at 0 °C; it was then

washed (pipette) with hexane and dried *in vacuo* to constant weight, to give the ketodiester B, m.p. 76.0–77.0 °C, yield 1.968 g (90.7%). Its NMR spectrum (in CCl_4) showed peaks at 2.2–3.1 (m, 6H, —CH_2—), 3.65 (s, 6H, —CO_2CH_3), 3.80 (s, 3H, —OCH_3) and 7.0–7.5 (m, 3H, arom. H) ppm.

Hydrolysis and decarboxylation of this (900 mg) by heating under reflux in nitrogen with acetic acid (5 ml), conc. HCl (1.25 ml) and water (1.0 ml) for 12 h, followed by removal of solvents *in vacuo* and recrystallization of the residue from carbon tetrachloride gave 1,2,3,4-tetrahydro-7-methoxy-1-oxo-naphthalene-2-acetic acid, m.p. 121–124 °C, in 87% yield.

Notes

1. Dimethyl carbonate was fractionated from CaH_2 and the fraction of b.p. 89–90 °C kept over basic alumina (act. I).

2. This azeotrope boils at 63 °C and contains 70% methanol.

3. The suspension thickens considerably but can still be stirred.

4. Diisopropyl ether is an excellent solvent for crystallization of low-melting polyfunctional compounds in that usually no oiling out on cooling is encountered. However, it readily forms peroxides; it should be tested frequently and is best kept under argon. Similar solvent properties are shown by carbon tetrachloride but crystallization from this is apt to be slow; the product may swim on top making mother liquor removal by pipette difficult; and it shows a tendency to form solvates.

II. Nucleophilic opening of an epoxide by a dianion

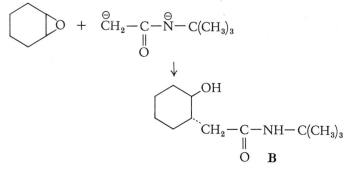

Apparatus: Two-necked 100 ml flask with magnetic stirring bar and combined cold-finger condenser and drying tube leading to nitrogen trap (Fig. 50).

Fig. 50

Cyclohexane (35 ml) was placed in the flask and first heated to reflux and then partly boiled off under nitrogen pressure (maintained subsequently) to a volume of 25 ml to remove traces of moisture. After cooling N,N'-tetramethylethylene-diamine (Notes 1 and 2, 1.05 ml, 0.82 g, 7 mmol) was added, and then N-t-butylacetamide[65] (Note 3, m.p. 98.5–99 °C, 755 mg, 6.6 mmol). The suspension was stirred, cooled to 5–10 °C (ice-water bath), and n-butyl lithium in hexane (1.59M, 8.2 ml, 13 mmol) added from a delivery burette (Fig. 19(c), p. 59) dropwise during 15 min. The suspension gave way to a turbid solution of the dianion. To this was added cyclohexene oxide (Note 4, 0.50 ml, 485 mg, 4.92 mmol), and the whole was heated under gentle reflux with stirring for 12 h. After cooling, acetic acid (0.9 ml) was added, followed by water (10 ml) and dichloromethane (5 ml). After stirring for 5 min the water layer was separated (pipette) and again extracted with ether–dichloromethane (2 : 1, 10 ml). The organic layers were washed with 1.5N sulfuric acid, with saturated sodium chloride solution, combined, dried and the solvents removed (rotary evaporator) to leave a viscous oil (1.42 g) which was transferred by dichloromethane to a three-bulb Kugelrohr tube. The excess of N-t-butylacetamide was sublimed into the far bulb at up to 100 °C (oven)/0.01 mm and the product then distilled into the near bulb at c. 150 °C/0.01 mm. The viscous gum obtained (943 mg) was transferred by boiling dichloromethane (p. 104) into a 25 ml Erlenmeyer flask and the solvent replaced by boiling diisopropyl ether to a final volume of 8 ml. On cooling after several hours the crystals were washed with cold diisopropyl ether and then hexane and dried to constant weight *in vacuo* to give the hydroxy-amide B, m.p. 102.5–103 °C, yield 860 mg (80.5%). Its infrared spectrum (in $CHCl_3$) showed bands at 3410 and 3310–3200 (OH, NH) and at 1664 (amide C=O) cm^{-1}; the NMR spectrum (in $CDCl_3$) showed peaks at 1.33 (s, 9H, —C(CH$_3$)$_3$), 3.19 (m, 1H, —CH—O), 3.98 (m, 1H, —OH, disappears with D_2O) and 5.75 (m, 1H, —NH—) ppm.

Notes

1. This amine was fractionated from sodium under nitrogen and the fraction of b.p. 121–122 °C kept over activated molecular sieves (5A) and CaH$_2$.

2. This solvent combination was chosen instead of tetrahydrofuran because of the latter's known instability to strong bases at and above room temperature.

3. The amide was dried overnight over P_2O_5. It was not dried by azeotropic distillation of its cyclohexane solution because of its known volatility and hence possible loss.

4. This was carefully fractionated to remove traces of water and the fraction of b.p. 129–130 °C kept over activated molecular sieves (5A).

III. Stereoselective alkylation of a nitrile

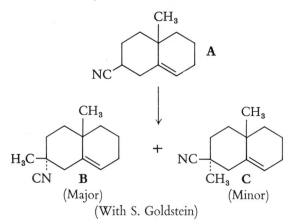

B
(Major)

C
(Minor)

(With S. Goldstein)

Apparatus: As in Fig. 49, except that the column was omitted and the reflux-distillation head mounted on the flask (of 100 ml capacity).

Benzene (25 ml) was placed in the flask, heated under reflux, and then distilled off to a volume of 15 ml under positive nitrogen pressure maintained subsequently. S was closed and T turned one-quarter to bypass the receiving flask, and the flask cooled to *c*. 10 °C. Diisopropylamine (Note 1, 1.00 ml, 714 mg, 7.05 mmol) was added and then *n*-butyl lithium in hexane (1.96M, 3.10 ml, 6.07 mmol). After stirring for 15 min at room temperature a benzene solution of the nitrile A[66] (m.p. 56.5–57.5 °C, 1.01 g, 5.77 mmol), which had been concentrated to a volume of 30 ml on the steam bath, was added (pipette) and washed in with another 3 ml dry benzene. The orange solution obtained was concentrated by distillation (S and T open) with stirring during 15 min to a volume of 15 ml (Note 2). It was then cooled to *c*. 5 °C (ice–water bath) (S closed, T turned one-quarter) and methyl iodide (Note 3, 2 ml, large excess) was added. After stirring briefly, the mixture was allowed to come to room temperature and then placed in a water bath (water level below inside level) at 40–45 °C for 3 h without stirring (Note 4). Cold water (10 ml) was added, the mixture stirred for

5 min, the aqueous layer removed (pipette) and again extracted (Erlenmeyer flask) with ether (10 ml). The organic layers were washed with 5% sodium sulfite solution, with saturated sodium chloride solution, combined, dried (Na_2SO_4) and the filtered solution concentrated (rotary evaporator). The residue, dissolved in hexane (5 ml), was applied to a column of basic alumina (act. II, 2 g) and washed through with more hexane to give 40 ml eluate. This was concentrated and the residue transferred in ether to a Kugelrohr tube. The mixture of nitriles B and C distilled at 90–100 °C (oven)/0.05 mm, with cooling of the receiving bulb by a methanol-soaked piece of cotton wool topped by a piece of dry ice. This gave 1.02 g of partly crystalline material which was shown by GLC (Carbowax 20M, 15% on Chromosorb W, 1 m by 6.35 mm, 213 °C) to contain B and C in the ratio of 4 : 1. The retention times of these, at 1 ml He s^{-1}, were found to be 9.0 and 10.0 min respectively; that of the starting material (only trace present) was 11.7 min. B and C were indistinguishable by TLC.

The distilled material was transferred by boiling pentane into a 15 ml Erlenmeyer flask and the solution concentrated to a volume of 5 ml. It was kept at 0 °C until crystallization proceeded and then at − 35 °C overnight. The flask was immersed in a dry ice–CCl_4 bath, the mother liquor was withdrawn and the crystals washed twice with small amounts of pentane cooled to − 35 °C. Drying *in vacuo* to constant weight gave the axial epimer B, m.p. 50–51 °C, shown by GLC to be >95% pure, yield 751 mg (68.8%). Its NMR spectrum (CCl_4) showed peaks at 1.05 (s, 3H, C—CH_3), 1.35 (s, 3H, C—CH_3), 2.18 (brd. S, 2H, allylic —CH_2—) and 5.55 (brd. t, 1H, =CH—) ppm. In epimer C the C-methyl peaks appeared at 1.13 and 1.27 p.p.m.

Notes

1. Diisopropylamine was fractionated from CaH_2 and the fraction of b.p. 83–83.5 °C kept over CaH_2.

2. Benzene does not appear to give an azeotrope with diisopropylamine, but the latter will co-distil with a relatively large amount of the solvent. If the deprotonation is not forced as

described, the yield of pure product is lower; and if tetrahydro-furan is used as solvent, the alkylation is less stereoselective.

3. Methyl iodide is best purified by passing it just before use through basic alumina (grade I).

4. There is no sense in stirring a solution which is homogeneous or where a precipitate formed (in this case lithium iodide) takes no further part in the reaction.

IV. (a) Reductive alkylation in liquid ammonia

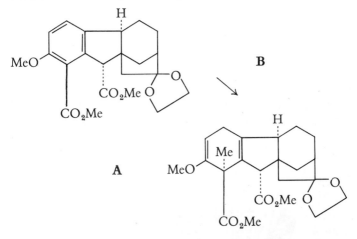

Apparatus: Two-necked 100 ml flask with magnetic stirring bar, thermometer adapter, small dry-ice condenser and calcium oxide drying tube connected via adapter to nitrogen trap (arrow), the last three items being connected by hooks and springs. The lower part of the condenser was later wrapped in gauze to catch condensation (Fig. 51).

The apparatus was carefully flamed out (without the thermometer) under positive nitrogen pressure (maintained throughout subsequently), and after closing and cooling, the low-temperature thermometer was inserted and the flask cooled to $-40\,°C$ to test dryness. The ketal-diester A[67] (600 mg, 1.49 mmol) was introduced and dissolved in dry tetrahydrofuran (8 ml) added directly from an alumina column (**Fig. 18, p. 56**). The flask

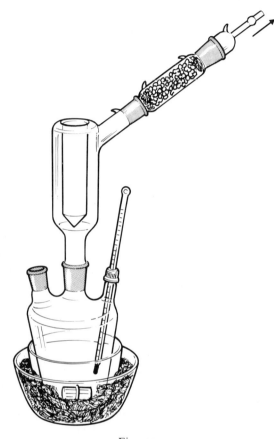

Fig. 51

was cooled in dry ice–methanol (coolant also placed in condenser); and dry ammonia gas, introduced into the side arm via a fitting adaptor (Note 1), was condensed in with continuous stirring to a total volume of 40 ml. Subsequently the reaction mixture was maintained at −60 to −70 °C (coolant level above inside level). With continuous even stirring to prevent splashing sodium was introduced in very small pieces (Note 2) until a deep blue colour (good illumination necessary) was permanent for

45 min. At this point methyl iodide (1 ml, large excess) was added leading to immediate discharge of colour. The stirred mixture was allowed to warm up to −35 °C, and solid sodium bisulfite (155 mg, 5 equiv., Note 3) was added. The ammonia was then allowed to evaporate through the nitrogen trap, finally by warming in a water bath at 30 °C. The flask was disconnected, the thermometer taken out, grease was removed, and the residual ammonia and solvent removed (rotary evaporator) at room temperature and 25 mm (some foaming evident at beginning).

To the solid residue was added water (20 ml) and chloroform containing 5% tetrahydrofuran (Note 4, 10 ml) and the whole stirred until two clear layers were formed. The organic layer was separated and the aqueous phase twice extracted with small portions of the same solvent. After washing with saturated NaCl solution and drying ($Na_2SO_4 + K_2CO_3$) the solvents were removed at room temperature and the residue treated with a small amount of ethereal diazomethane to esterify any acidic material present. It was then taken up in benzene and the solution was passed through basic alumina (act. II, 2.5 g) and washed through with benzene. After solvent removal the product was recrystallized from dichloromethane–diisopropyl ether–hexane to give the enol ether B, m.p. 150–155 °C, suitable for the next step, yield 382 mg (61%). The analytical sample had m.p. 156–156.5 °C and its NMR spectrum ($CDCl_3$) showed peaks at 1.47 (s, 3H, C—CH_3), 3.50 (s, 6H, —CO_2CH_3), 3.55 (s, 4H, ketal —CH_2—), 3.78 (s, 3H, —OCH_3) and 4.70 (t, 1H, =CH—) ppm.

Notes

1. Ammonia from a cylinder was dried by passing it through two drying towers (KOH and CaO).

2. A 100 mg piece of clean sodium was weighed out in a small beaker under petroleum ether (b.p. 100–120 °C); this was carefully heated on a hotplate until the sodium had melted. It was then stirred with a micro-spatula so that on cooling it solidified in globules of 2–3 mm diameter. These were added by

impaling on the tip of a scalpel and stripping off with a pair of forceps directly into the reaction mixture. Approximately 75 mg (2.1 equiv.) were required.

3. This adjusts the resulting aqueous solution to pH 6 and prevents complications due to iodine formed by oxidation.

4. The ether solvent (a Lewis base) was added in view of the extreme sensitivity of the product to traces of mineral acid usually present in chloroform.

IV. (b) Selective cleavage of an enol ether, and sodium borohydride reduction of the product

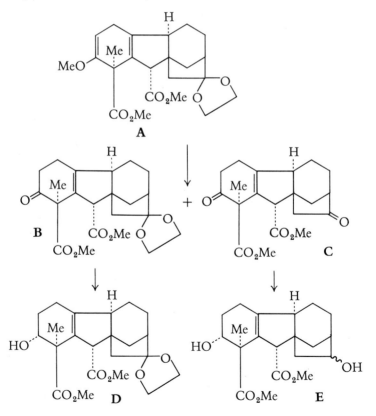

In a 100 ml two-necked flask with magnetic stirring bar and drying tube connected to nitrogen the above product (418 mg, 1.00 mmol) was dissolved in dry benzene (40 ml). To this was added with stirring 2.2 ml of a benzene solution which was 1.08M in trichloracetic acid and 0.5M in water (Note 1). The reaction was followed at room temperature (20–23 °C) by TLC (alumina plates, ethyl acetate–cyclohexane 1 : 8, two developments). The least polar spot of the starting material A gave way to a more polar spot of B, and finally (after 3.5 h) a still more polar spot (due to C) began to form. At this point a solution of sodium bicarbonate (300 mg) in water (6 ml) was added with stirring, the benzene layer was washed with potassium carbonate solution (5%), dried (Na_2SO_4) and the solvent removed, finally at 40 °C/0.1 mm (Note 2).

The residue was dissolved in methanol (5 ml) and tetrahydrofuran (2.5 ml), and sodium borohydride (60 mg, Note 3) was added and the whole stirred briefly, at 0 °C. The solution was allowed to reach room temperature overnight, the solvents were removed (rotary evaporator) at room temperature, water was added and the product extracted several times with small quantities of dichloromethane. Drying (Na_2SO_4) and solvent removal gave a residue (501 mg) which was chromatographed on alumina (basic, grade II, 10 g) starting with dichloromethane–hexane (1 : 2) and eluting with increasing proportions of dichloromethane. At first unchanged starting material A (26 mg) was eluted, followed by the hydroxyketal diester (D) (371 mg), which after recrystallization from dichloromethane–diisopropyl ether gave 301 mg (66.5%) of m.p. 141–142.5 °C. The NMR spectrum of the derived O-acetate (m.p. 116.5–117.5 °C) (in $CDCl_3$) showed (*inter alia*) peaks at 1.30 (s, 3H, C—CH_3), 2.07 (s, 3H, —$COCH_3$), 3.65 (s, 6H, —CO_2CH_3), 3.90 (brd. s, 4H, ketal —CH_2—) and 4.8–5.1 (m, 1H, axial —CH—OAc) ppm.

Final elution with chloroform gave a mixture of diols E (41 mg).

Notes

1. This was prepared by azeotropically drying a benzene

solution of the appropriate amount of trichloroacetic acid (puriss. grade), adding the required amount of water and making up to required volume with dry benzene. Its exact acid content was determined by adding an aliquot to water and titrating against standard alkali.

2. The products were not separated by chromatography at this stage because of the greater polarity difference between the products of the next step.

3. A considerable excess of sodium borohydride was used in view of its known instability in methanol; the latter solvent was chosen instead of ethanol (in which it is more stable) to avoid possible transesterification.

8
Various Hints and Gadgets

SMALL-SCALE CONDENSATION REACTIONS

These include Knoevenagel condensations, enamine formation, acetalization, etc. The customary Dean–Stark traps, especially if of the size commercially available, are far too lengthy even for work with larger quantities. On a millimole scale there is of course no point in even trying to measure the amount of water evolved in a condensation reaction, but the problem of trapping the water evolved efficiently and continuously still remains. A useful adapter for such cases is shown in Fig. 52. At the bottom of the bulb one can place a small bed of self-indicating silica gel, or (with basic substrates and catalysts) of activated molecular sieves.

OIL-JACKETED DISTILLATION FLASKS

These, as exemplified in Fig. 53, are very suitable for distillation of high-boiling and sensitive materials in a high vacuum, in

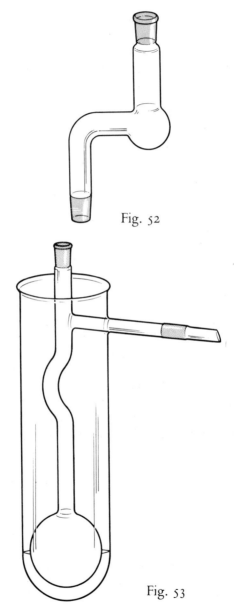

Fig. 52

Fig. 53

particular where quantities beyond the range of a Kugelrohr tube and oven are involved. The outer jacket can be filled with the heating oil practically to the top, but allowance has to be made for expansion during heating. For added stability the distillation bulb must be fused to the jacket in several places, and proper annealing of the whole is essential. The material is introduced in principle as with a Kugelrohr tube: from the top by pipette and with the apparatus inclined so that nothing comes out via the side arm and with the whole immersed in a steam bath to heat the oil sufficiently to drive off the low-boiling solvent used for transfer.

These were first seen by the author in Professor G. Stork's laboratory (Columbia University, 1954), but apologies are extended to anyone claiming prior rights. They are not apparently commercially available.

SHAKING

Should you come across an ancient sewing machine, of the foot-operated type (junk dealer, grandmother's attic), requisition it. Unfortunately these are becoming collector's items. The treadle, when connected to an electric motor, is a very suitable support for items to be shaken, and the open lattice work usually encountered makes firm attachment by springs or rubber bands an easy matter. To be on the safe side the treadle should be fitted with a rim to prevent items from sliding off.

BURETTE FOR 'JONES' OXIDATIONS

In this method of oxidation (generally of a secondary alcohol to a ketone) by titrating in acetone solution with 8N chromic acid in dilute sulfuric acid the volume of reagent required is usually so small that on a scale less than 1 mmol the drop size obtained from a regular burette tip is too large for accurate work. It will then be necessary to use a micro-tip adapter like the one shown in Fig. 54, which incorporates a capillary made of a resistant metal which can deliver individual drops of 0.01 ml volume. These, with the interfitting micro-burette, are supplied, for example, by Kimble Division, Owens Illinois Inc. (Cat. No. 17100–F). The adapter has to be held in place by a small rubber band. An air bubble inevitably remains in the adapter bulb but its volume usually stays constant.

Fig. 54

SUBLIMATION AND TRANSFER OF POTASSIUM *T*-BUTOXIDE AND SIMILAR SENSITIVE MATERIALS

This material is available commercially from a number of sources, but one can never be quite sure of its purity, or whether part of the material may not in fact be the 1 : 1 complex with *t*-butanol. When it is used in stoichiometric amounts, as is often the case, the only way to purify it is by sublimation in a high vacuum. The customary large-scale sublimation apparatus is unsuitable, because it is too large to introduce into a drybox in which it will be necessary to handle, transfer and bottle the material. This problem can be solved by doing the sublimation in a large (50 mm o.d.) and relatively short test-tube (Fig. 55) which will just about fit into the standard Kugelrohr oven. It is divided by a thick wedge into two unequal compartments, and a short stopcock inside a well-fitting rubber stopper leads to the vacuum system. In such a tube quantities of up to 25 g of crude material can be sublimed at one time; cooling is not generally necessary. If the highest possible vacuum is used (better than 0.05 mm Hg) and the sublimation conducted at the lowest feasible temperature (170–180 °C) the sublimate should collect as compact crystalline scales which can easily be scraped off, lifted out and transferred into a bottle with a suitable spatula, all inside the drybox, while naturally the tube is kept horizontal to prevent contamination with the sublimation residue.

Fig. 55

PREPARING SUSPENSIONS FOR COATING TLC PLATES

Experience has shown that thorough mixing of such suspensions and shortest possible mixing time will together ensure complete homogeneity and avoid any chemical interaction with the liquid medium which would lead to formation of air bubbles. This is where an ultrasonic cleaning bath comes in useful (Moshe Weisz, School of Pharmacy, Hebrew University, Jerusalem). Thus, a container (jam jar with cover) with a mixture of 100 g silica gel PF_{254} in 240 ml water is shaken first manually for 30–60 s, then in the half-filled ultrasonic bath for 4–5 min, allowed to stand for 2 min, again shaken manually for 30 s and then again ultrasonically for 2 min, after which it is ready for immediate spreading. Such suspensions have always given results superior to those prepared according to manufacturer's instructions.

THE SCHREIBER ADAPTER

This (see Fig. 56), as suggested by J. Schreiber (E.T.H., Zürich), is a useful item when a reaction has to be conducted under inert gas in a one-necked flask. Reactants can be added from the top while inert gas flushing is done through the side arm.

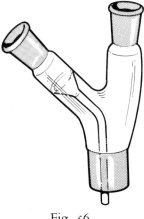

Fig. 56

CUTTING SMALL FILTER PAPER CIRCLES

Often it is impossible to buy sizes to fit a particular Hirsch funnel. A good tool for cutting them out of larger sizes is a very sharp and scrupulously clean cork borer with the appropriate diameter.

VACUUM REDISTILLATION OF SMALL AMOUNTS DIRECTLY INTO A VIAL

If you are faced with this necessity, particularly with viscous liquids which cannot be quantitatively transferred from a

Kugelrohr tube, the apparatus shown in Fig. 57, due to Niels Clauson-Kaas (Copenhagen)[68], can be used. The liquid to be distilled is introduced into the distilling bulb (kept on the steam bath) by capillary pipette. Vacuum is applied at T; the vial V rests on a standard joint glass stopper (hooked on by springs) as shown.

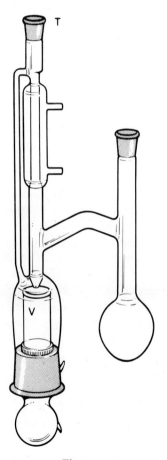

Fig. 57

CLEANING SODIUM

Doing this with a knife is not only wasteful but leads to a dangerous accumulation of sodium residues. One way of purification[69] is to melt carefully under dry xylene and allow the liquid metal to flow out of the 'shell' of surrounding hydroxide. In a second method[70] the metal, where available in spherical pieces, is 'descaled' by shaking inside a cylindrical metal rasp under petroleum ether. Another possibility is to shave it with a corn shaver (Fig. 58) of the type that utilizes pieces of razor blade, while for larger pieces the use of a sharpened potato peeler might not be inappropriate.

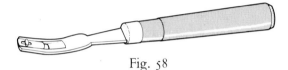

Fig. 58

STORING AND CLOSING KUGELROHR BULBS

These are awkward to hold by clamp; and for storing them in the refrigerator one rarely succeeds in finding an object of the right shape and size (e.g. shallow beaker) which will hold them securely without risk of upsetting and spilling the contents. A simple way out of this dilemma is to use a shallow wooden box (e.g. cigar box) as shown in Fig. 59, with rubber bands stretched around it at regular intervals; on these the bulbs are held quite securely. A good way of closing them at both ends is to slip on the inner (hollow) part of a serum rubber stopper (Fig. 60).

TEFLON® RING STANDARD JOINT CONNECTORS

Who has not at some time felt exasperation and annoyance at the way in which many commercially available adapters lead to unnecessary elongation and troublesome constriction? I strongly believe that whoever has become used instead to the use of Teflon® adapter rings (Fig. 61) will never use a glass

147

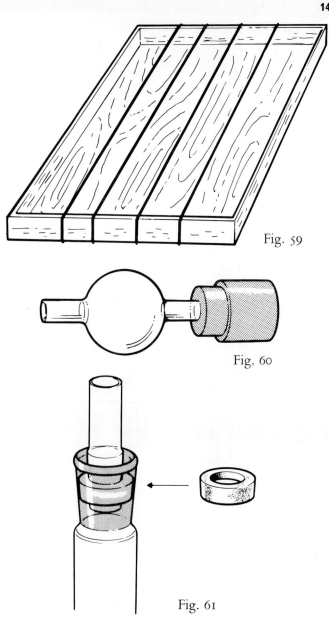

Fig. 59

Fig. 60

Fig. 61

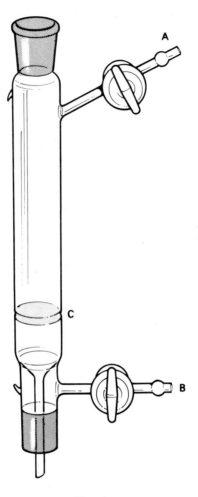

Fig. 62

adapter again. These are now commercially available, and if not they can be turned on a precision lathe by a reasonably skilled operator. Their height need not be greater than as shown in Fig. 61.

FILTRATION IN AN INERT ATMOSPHERE

The adapter illustrated in Fig. 62, due to M. Cais (Technion, Haifa), is very handy for filtering a suspension introduced from a reaction vessel under inert gas pressure and while maintaining an inert atmosphere throughout. Filtration can be by suction applied at B, or under pressure from A with the top stopper securely fastened down. Obviously when using this the multi-outlet nitrogen trap (p. 42) is very appropriate. The sintered glass disc (C) should generally be of medium (No. 2 or 3) porosity. If the object is to collect an air-sensitive solid this can be dried *in vacuo* and then handled under inert gas. As with all glassware incorporating sintered glass, this needs very thorough cleaning and drying after use.

9
On Ordering, Bottling and Storing Chemicals

The time has come for some kind of Consumer's Guide on chemicals for the research worker. For obvious reasons no names of individual suppliers will be mentioned except perhaps where praise is due. However, the number of chemical supply firms worldwide is relatively limited and readers with some experience will not have much difficulty in reading between the lines.

I realize that some of the following remarks might be more relevant to people other than those to whom this book is primarily addressed, but they had to be written, and if not here —where else?

CHOOSING A SUPPLIER

On the whole the best pointer is the supplier's catalogue. Naturally this cannot contain all possibly desirable information but some suppliers do try their best. Some questions that should be raised in every case include the following:

1. To what extent does the catalogue contain full information, not only on prices but also on purity, physical constants and packaging of chemicals?

2. Does it cross-reference alternative names of products; and are there a Formula Index and a list explaining without exception all abbreviations used?

3. Does it bear a definite date of issue and not merely a serial number?

4. How frequently have complete issues come out in the past; and how often has the supplier sent out lists indicating changes (changed prices, new package sizes, new and cancelled items), and all this without having had to be prompted?

5. How well is the catalogue organized with the express purpose of helping the reader and customer?

It is my experience that answers to questions such as these are quite a reliable pointer to one's eventual experience with the supplier—his reliability, his helpfulness to the customer and the quality of the product he sells. To this must be added the favorable impression created by advertizing which does not just push the product but gives helpful information and references on its use, not to speak of the supplier who is sufficiently interested in his customer to publish, at his own expense, interesting and useful information which has no direct connection with his product at all.

The large suppliers manufacture only a part of the products which they list. The rest originate with smaller subcontracted firms—and there may be many of those. All the same, it is the direct supplier who is to be held accountable for the product and for detailed information on it.

These days it may happen quite frequently that a company is taken over by another, and this may cause trouble and aggravation. While such an often prolonged process is taking place it may well be the customer whose interests are pushed to the bottom of the priority list, to the extent that not even registered letters and telegrams are answered. If and when you get an inkling of such a situation try to deal with a subcontracted company directly or, better still, quickly change over to another supplier and stick with him even after the former company, under new ownership, have decided once again to be aware of your existence. As far as you are concerned they are a new company who will have to build up their reputation from scratch.

For out-of-the-ordinary chemicals not listed by the larger firms *Chem Sources*[71] can be a useful guide. However, when enquiring about such a material with a lesser known firm your first question should be on a firm delivery date. Frequently it will be out of stock and the next batch will be made only after enough interest as measured by the number of enquiries has accumulated. By that time the chances are that you will have forgotten what you wanted it for. If immediate delivery is not promised it may be better (and cheaper for certain) to make it yourself.

Never hesitate to ask for information and samples offered in advertisements in journals such as *Chemical and Engineering News*, *Angewandte Chemie* and *Chemistry in Britain*. What you will get in return is always worth more than the effort or even the cost of the stamp.

QUALITY AND PRICE

Ordering a starting material, reagent or solvent is by no means a matter of comparing catalogue prices alone. Even weighing up price against (alleged) quality should not lead to a final conclusion on from whom to buy.

On the subject of quality: individual firms' ideas on terms like 'Reagent Quality', 'Pure' and even 'Puriss.' can be quite elastic. One company's 'Chemically Pure' may be no different from another's 'Technical Grade'. Even a stated batch analysis must be treated with reservation. In nearly every conceivable situation you will be buying not a whole batch but one or two bottles from that batch. You have no way of finding out about the personal habits of the man who filled your bottle or whether the bottling machine was clean and in full working order that day. Furthermore, a sensitive product may spoil because of faulty packaging (see below) even by the time it reaches you. Finally, even the most reputable suppliers have been known to supply bottles whose contents bore little if any relation to the label. True, 'to err is human', but neither the profusest apology nor the supplier's willingness to replace free of charge can ever make up for time and effort lost. Your Golden Rule should be: If in any doubt, and even if you think you have no doubt, redistil, recrystallize and resublime whenever possible; and with products that cannot be purified: be on guard twice over! And you should be equally skeptical when faced with published work that states that 'Material X was used as received from the Y company'.

One should always keep in mind that what matters in the majority of organic starting materials and reagents is not the price per gram (or 100 g) but the price per gram/mole. This is of particular importance when a choice has to be made between alternative reagents for a specific purpose and where the advantage of a particular one has not been established beyond doubt. Some examples are: radical halogenation agents, hindered secondary or tertiary strong amine bases, reagents for forming peptide bonds or for introducing double bonds.

PACKAGING AND BOTTLING

It is astonishing how often even experienced research workers, after having devoted much effort and thought to the preparation of a pure product, become careless and thoughtless when it comes to bottling and storing it. This goes to an even larger extent for manufacturers and suppliers.

It can be said without hesitation that for many chemicals, especially when corrosive (aluminum chloride, phosphorus halides), volatile (amines), or generally dangerous on exposure (alkali metals and hydrides), the question of packaging can far outweigh the cost factor when choosing a source of supply. Add to that the safety factor: a serious plane crash some years ago was strongly suspected to be caused by improper packaging of a corrosive chemical.[72]

Glass-stoppered bottles are unsuitable for any purpose except shelf storage of limited amounts of organic solvents or of mineral acids. They show no resistance to internal pressure build-up, and on the other hand the stopper will inevitably freeze if the bottle contains a solution of a solid or if the contents are liable to solidify on exposure. Greasing achieves little beyond contaminating the contents. Polyethylene or polypropylene bottles are very suitable for storage of most inorganic solids or of neutral or alkaline inorganic solutions. Cardboard or aluminum foil-lined screw-capped bottles still used routinely by a number of firms do not sufficiently protect the contents in many cases. Those whose closure is lined by a Teflon® film are satisfactory but only so long as this lining is intact and not folded or broken. There is nothing to be said against closure of the crown cork type except on the little problem of what to do after they have once been opened.

Probably the best type of container is the screw-capped glass bottle whose cap has a built-in plastic cone (used by a number of U.S. firms), or better still which incorporate a plastic insert which fits snugly into the neck and whose lip is pressed down tightly by the screw-cap (used, for example, by the Koch-Light Co.). In my experience bottles closed in this way can be used for storage for a year or longer of such sensitive materials as trifluoroacetic anhydride, titanium tetrachloride, chlorosulfonyl isocyanate, volatile amines and even alkyllithium solutions. The inserts can be purchased separately and very cheaply in a large range of sizes. I have come across only one case where this closure was attacked: a solution of diethylaluminum chloride, and here the problem was solved by having a similar Teflon insert machined specially.

Aldrich Chemical Co. have very recently introduced their rather elaborate Sure/Seal type of bottle closure. This is designed for withdrawal by syringe, with the difference that there are *two* elastomer liners (each backed with Teflon®), the idea being that the second will serve to seal the puncture hole in the first.

While on the subject of packaging chemicals one cannot but single out Merck and Co. for special praise. The distinctive types of closure used by this firm are obviously the result of much thought on this subject. And this company, incidentally, appear to be the only one who take the trouble of indicating the types of packaging used in each case in their catalogue.

Ampoules

One should ever keep in mind that once opened these can never be satisfactorily reclosed except by resealing at (usually) considerable risk. Of course there are chemicals that cannot be packaged in any other form, and here it will pay in the long run to order only the smallest size available on the market, or to arrange ahead of time for immediate transfer of the contents.

The Beer Can Problem

Beware of chemicals supplied in sealed metal tins which, once opened, can never be reclosed, and this refers particularly to the now-popular snap-open type. With these you will often have nasty surprises. For example:

(a) finding lithium aluminium hydride in loose form which leads to a mad scramble for a suitable dry and well-closed container;

(b) encountering a solid block of sodium or even (believe it or not) potassium which nobody can be expected to cut down to size unless suicidally inclined; or

(c) a soup of lithium dispersion with the finely divided and exceedingly active metal particles swimming on top.

Such items usually come to be ordered because they look 'cheaper' from the catalogue. The supplier's label rarely prepares you for what is in store inside, and there ought to be a law against such Jacks in Boxes if there is not one on the books already. If you have any indication that the goods you order will come 'canned' demand full details in advance and in writing.

ORDERING ALKALI METALS AND HYDRIDES

This requires special thought and attention. Alkali metals (sodium, potassium, lithium) should never be ordered dry in hermetically sealed cans unless you are sure beforehand that you will use the entire contents at once. This happens in industrial establishments but probably never in a research laboratory. They should be ordered only if protected by mineral oil or grease and this should be ascertained in advance if the catalogue is not clear on this point. There remains the question of particle size. Material in wire form (e.g. lithium) is very convenient for small-scale work but will inevitably spoil by surface oxidation after a time. With larger pieces there is less risk of this, but cutting them to size is always apt to be hazardous. It is probably best to compromise by ordering in 1–3 g pieces or sticks, such as sodium supplied by B.D.H. under paraffin oil, which comes in very suitable screw-lidded metal cans incorporating a rubber seal.

It is well known that the reactivity of lithium is quite dependent on the amount of sodium it contains. Any supplier who is unable or unwilling to supply exact information on this point should be passed over; and lithium is not the only example where this policy should be followed.

Suspensions and dispersions of alkali metals, hydrides and amides are a special and vexing problem. The advantage arising from their great reactivity are usually offset by not knowing just how homogeneous they are and hence how accurately they can be dispensed. Catalogues are never clear on this point; and here again it should be the supplier's responsibility, and not that of subcontracted firms from which he obtained the material in the first place, to give you full and exact information.

METHYL AND OTHER ESTERS

When ordering and utilizing one of the common reactive esters (malonates, acetoacetates, cyanoacetates, haloacetates, etc.) there is rarely any sense in choosing the ethyl or higher alkyl ester if the methyl ester is available; and even if it is not it usually pays to make it. Methyl esters are lower boiling and at the same time more likely to be crystalline. Another important advantage is the simplicity of the NMR spectrum. Moreover, when the ester group has to be reintroduced at a later stage diazomethane is much more easily available than other diazoalkanes. On the whole, the price of methyl esters on a molar basis is only slightly higher than that of the ethyl ester and may occasionally be lower. The disadvantage of having to use the lower-boiling methanol in, for example, a condensation reaction, is more than offset by the very much higher solubility of alkali methoxides in this solvent compared with that of the ethoxides in ethanol, quite apart from the greater ease of making absolute methanol.

These considerations should apply also to most acetals, ortho-esters, ethers and other functional groups where the nature of the alkyl group is not of intrinsic importance (as it certainly is in the case of *t*-butyl or benzyl).

ADSORBENTS

It is absolutely vital to order these only from reputable suppliers who are willing to stake their reputation at the very least by giving full and checkable details—mesh size, Brockmann activity, pH—on the label. Simple calculation will show what happens on chromatographing 1 g of product on 200 g of adsorbent of doubtful origin which may contain 'merely' 0.5% of a probably soluble contaminant; and of course there is no way to 'purify' such a material. In choosing a supplier it is best to consult other workers with wide and current experience in chromatographic work, and to check the batch number of material you are getting compared with that you will find in other laboratories. In this field one takes a chance at one's peril.

ORDERING, REBOTTLING AND TRANSFER OF SENSITIVE LIQUIDS AND SOLUTIONS

As an example let us take *n*-butyl lithium, among the most widely used reagents at present, usually in hydrocarbon solution (hexane, heptane, benzene). Considerable savings can be made if this is ordered in larger quantities, for example, the half- or one-liter size sold by Merck and Co. Certain U.S. suppliers also sell this in rubber septum-closed bottles, on this subject see p. 62.

However, each time such a bottle is opened (and the bigger it is the more often) spoilage by air or moisture is almost unavoidable, even when stringent precautions are taken and all the more when, as is usually the case, such a bottle is 'shared' by a number of people. The best way to solve this problem is to transfer the contents on arrival to 50 ml or 100 ml screw-capped plastic insert bottles as mentioned above, using the simple device illustrated in Fig. 21(a), or that in Fig. 21(b) (p. 61) if you are faced with a bottle into which one cannot even insert a 25 ml graduated pipette.

Certain reputable suppliers can also offer adsorbents in large package sizes; and these too should be rebottled, by an experienced worker and on a dry day.

It must be kept in mind that quotations for reagents in such package sizes do not usually appear in a regular catalogue, and they should be asked for by mail.

SOLVENTS

This is usually the biggest single item in any research budget, especially where much chromatographic work is done (not to speak of washing glassware), and even where provision for partial recovery is made.

The following figures are calculated on the basis of actual experience for a group of 50–60 laboratory workers. Prices are in U.S. $ for convenience only:

Solvent	Annual consumption/ kg	Total price, techn. or pract. in drums	Total price[a] anal. or puriss.	Saving
Acetone	134[b]	72	346	274
Benzene	404	302	1232	912
Chloroform	1034	1008	3774	2 766
Ethanol	514	622[c]	1502[c, e]	880
Ethyl acetate	404	280	1176	896
Hexane	364	646[d]	4282	3 636
Methanol	480	146	2030[e]	1 884
Methylene chloride	1416	914	3158	2 244
			Total savings: (or 77%)	$13 492

[a] In 2.5 liter bottles after maximum quantity discount, e.g. 36 bottles bought at same time

[b] Has been phased out for cleaning, used for chemical purposes only.

[c] Not including duty.

[d] 'Hexane fraction'.

[e] 'Absolute'.

From these figures for just eight of the common solvents (ether is not included because of the special problems involved) it is obvious that what can be saved by bulk buying of solvents of practical or technical grade can go a long way toward covering the salary of a technician employed primarily on purification and redistillation and perhaps even to cover part of the overhead costs. Also such a technician would be in the best position to organize distribution among groups, prevent accumulation of bottles in laboratories and keep an eye on wastage. And, most importantly, research workers can be pretty sure of, and have direct control over, what they are getting. Last but not least there is the aspect of storage and hence safety. The main customers for solvents in bulk and in drums are other industrial firms. These are far more demanding in regard to packaging free of corrosion problems and safe for prolonged storage than the average university or research administration, and thus they will demand and get a lot more attention in this regard from the supplier.

ORDERING CHEMICALS BY AIR MAIL AND AIR FREIGHT

When as frequently happens a chemical must be ordered urgently by air the first thing to investigate is if and how its transportation is permitted by this route. Here common sense and The Law do not necessarily coincide, and it is the latter which must be followed as laid down by the International Air Transport Association (IATA) in their 'Restricted Articles Regulation'.[73] Most shippers look upon these as Holy Writ, the same as most (but by no means all) travel agents regard the 'ABC' and the 'APT', and will go strictly by what appears and does not appear in print. Hence it is a good thing to have access to a copy of these Regulations and to study them carefully.

As for getting chemicals through customs quickly a poison label often works wonders and will in principle never be far from the truth. A certain venerable German chemical firm used to put a poison label even on a bottle of sodium chloride. And if you should receive a thus-marked chemical from a supplier in Germany, Austria or Switzerland you could always argue at a pinch that it was sent as a present. Labels marked 'Radioactive' are even better, but that would really be cheating and could get the supplier (and even you) into serious trouble.

While on this subject: customs authorities anywhere should never be treated with more respect than they deserve. A concrete example: payment was demanded on a cylinder of helium (normally duty-free in this case) stated to contain 0.95% of isobutane (Geiger Gas), on the grounds that 'the book says that isobutane is dutiable'. The problem was solved to everyone's satisfaction by having future cylinders shipped as 'Helium, 99.05% Pure'.

LABELLING AND STORAGE

Everything should be done to prevent a bottle from becoming a dead loss, through fading of the writing or lettering on the label or its detachment. This happens frequently even with labels used by commercial firms. Old-fashioned black ink lasts a long time and in my experience can still be legible after more than 25 years. Various makes of ball pens differ widely in light fastness and resistance to chemical vapors; on this it pays to get an expert opinion, for example from someone working in an archival department. But, as an industrial colleague[74] has kindly pointed out, the most permanent label is that written by the common graphite pencil—as is borne out by his experience and as should be self-evident to any chemist (provided the label is not subjected to any mechanical abrasion in the course of time).

Labels should be covered by a glaze of the cellulose acetate type. Transparent adhesive tape is strongly advised against because in the course of time the ink may diffuse into the plastic layer and the writing will become illegible. *Never* use chinagraph pencils for permanent labelling.

All bottles should show the name, structural formula, melting or boiling point and molecular weight of the contents, and date of receipt (or preparation). Bottles containing liquids should also show the specific gravity. Among commercial suppliers Fluka AG merit special praise for the amount of information given on their labels and that includes gas-chromatographically checked purity data on practically every liquid product sold by them; against that there are still a number of firms (including one major supplier) who lack this elementary courtesy towards the customer. What is of course even worse is information that is actually misleading. For example: solutions stated to be '15%' without saying whether this is w/w or w/v, or without giving the specific gravity even with solutions of products proclaimed as 'spontaneously inflammable' and thus highly dangerous to weigh out. Or solutions described as 'about 20%': plus or minus how much? (since you are paying for it!). More people should write and kick up a fuss about this sort of practice; perhaps we may then arrive at long last at a situation where all concentrations are described in terms of *molarity* which is the most useful and meaningful one.

Another dead loss is the bottle which cannot be found. As many chemicals used within a group as possible should be stored centrally. Organic compounds should preferably be in order of functional groups, and inorganic ones by element or alphabetically. If this is controlled by one person who keeps in permanent touch with other groups, and using a card index, a lot of aggravation can be avoided and a lot of money saved. Any research worker with some years of experience knows how much money is wasted through purchase of a chemical (or, for that matter, any piece of equipment) which was there all the time in some corner, unused and unknown until it came to light at the next spring cleaning.

On the subject of storage location the main consideration should be the safety one. Alkali metals and hydrides should be kept as far as at all possible from work benches and water faucets—best of all in a cupboard or wall shelf. Bottles containing corrosive materials should be placed in plastic containers, preferably together with inert absorptive material—to protect both neighbouring bottles and the shelf they are standing on. Solvents should be stored as low as possible, under the hood or better still in closed cupboards,* but definitely not on the floor or anywhere near a source of heat. And the best place to keep catalysts is under lock and key and in an office room.

---

* It is now increasingly believed by laboratory planners that the advantage arising from superior heat insulation on storing flammable material in wooden cupboards[75] can outweigh the apparent disadvantages, particularly now that fire-proofing wood is a process that is both simple and effective.

References

1. E. J. Crane, A. M. Patterson and E. B. Marr, *A Guide to the Literature of Chemistry*, 2nd Edn, Wiley, New York/Chapman and Hall, London, 1957.
2. G. M. Dyson, *A Short Guide to Chemical Literature*, Longmans Green, London, 1951.
3. R. T. Bottle (Ed.), *Use of the Chemical Literature*, Butterworth, London, 1962.
4. C. R. Burman, *How to Find Out in Chemistry*, 2nd Edn, Pergamon, Oxford, 1966.
5. *Searching the Chemical Literature*, Advances in Chemistry Series No. 30, American Chemical Society, Washington, D.C., 1961.
6. J. E. Hendrickson, D. J. Cram and G. S. Hammond, *Organic Chemistry*, 3rd Edn, McGraw-Hill, New York, 1970, p. 1140.
7. T. C. Owen and R. M. W. Rickett, Beilstein's Handbuch as a Source of Information on Organic Chemistry, in *Use of the Chemical Literature* (R. T. Bottle, Ed.), Butterworth, London, 1962, p. 130.
8. O. Weissbach, *The Beilstein Guide*, Springer-Verlag, Berlin, 1976.
9. *Elsevier's Encyclopedia of Organic Chemistry*, Ed. F. Rath, Elsevier, Amsterdam/Springer-Verlag, Berlin, 1948–62.
10. J. Jaques, H. Kagan and G. Ourisson, *Pouvoir Rotatoire Naturel. 1a. Stéroïdes*, Pergamon, Oxford, 1965.

164   **GUIDE FOR THE ORGANIC EXPERIMENTALIST**

11. W. Karrer, *Konstitution und Vorkommen der Organischen Pflanzenstoffe*, Birkhaeuser Verlag, Basel, 1958.

12. W. Theilheimer (Ed.), *Synthetic Methods of Organic Chemistry*, Karger, Basel, Vols 1–31, 1948–77.

13. *Houben–Weyl's Methoden der Organischen Chemie* (E. Mueller, Ed.), Georg Thieme Verlag, Stuttgart, 48 Vols (from 1953).

14. *Technique of Organic Chemistry* (A. Weissberger, Ed.), Vols I–XIV, Interscience, New York, 1951–69.

15. *Techniques of Chemistry* (A. Weissberger, Ed.), Vols I-XI, Wiley–Interscience, New York, 1970–76.

16. *Organic Reactions*, Wiley, New York, Vols 1–25 (from 1942).

17. L. F. Fieser and M. Fieser, *Reagents for Organic Synthesis*, Wiley, New York, Vols 1–6 (from 1967).

18. *Annual Reports in Organic Synthesis* (J. McMurray, R. B. Miller, L. S. Hegedus and S. R. Wilson, Eds), Academic Press, New York, from 1970.

19. I. T. Harrison and S. Harrison, *Compendium of Organic Synthetic Methods*, Wiley–Interscience, New York, 2 Vols, 1974.

20. *Organic Reactions in Steroid Chemistry* (J. Fried and J. A. Edwards, Eds) 2 Vols, van Nostrand–Reinhold, New York, 1972.

21. *Steroid Reactions* (C. Djerassi, Ed.), Holden-Day, San Francisco, 1963.

22. H. J. E. Loewenthal, 'Selective Reactions and Modification of Functional Groups in Steroid Chemistry', *Tetrahedron*, **6**, 269 (1959).

23. H. O. House, *Modern Synthetic Reactions*, 2nd Edn, Benjamin, New York, 1972.

24. S. R. Sandler and W. Caro, *Organic Functional Group Preparations*, 3 Vols, Academic Press, New York, from 1968.

25. R. B. Wagner and H. D. Zook, *Synthetic Organic Chemistry*, Wiley, New York/Chapman and Hall, London, 1953.

26. *Friedel–Crafts and Related Reactions* (G. A. Olah, Ed.), Vols I–IV, Interscience, New York, 1963–65.

27. J. Mathieu and J. Weill-Raynal, *Formation of C—C Bonds*, George Thieme Verlag, Stuttgart. Vol. I: Introduction of a Functional Carbon Atom (1973); Vol. II: Introduction of a Carbon Chain or Aromatic Ring (1975).

28. *Organic Syntheses*, Collective Vols 1–5, Vols 50–56, Wiley, New York, 1932–77.

29. D. H. Lewis, *Index of Reviews in Organic Chemistry*, Cumulative Issue, The Chemical Society, London, 1976 (and previous issues).

30. *Organic Reaction Mechanisms* (B. Capon, C. W. Rees, A. R. Butler and M. J. Perkins, Eds), Wiley, New York, annual since 1965.

31. E. Garfield, *ChemTech*, **6**, 167 (1976); *Current Abstracts in Chemistry and Index Chemicus*, **62**, No. 9 (1976).

32. E. F. Eikenberry, *J. Chem. Educ.* **52**, 385 (1975).

33. A. J. Gordon and R. A. Ford, *The Chemist's Companion*, Wiley, New York, 1972, p. 449.

34. A. J. Gordon and R. A. Ford, *The Chemist's Companion*, Wiley, New York, 1972, p. 451.

35. W. S. Johnson and W. P. Schneider, *Organic Syntheses*, Coll. Vol. 4, Wiley, New York, 1963, p. 132.

36. Woelm AG, Information Sheet No. 3; see also *Drying in the Laboratory*, Merck and Co.

37. S. G. Watson and J. F. Eastham, *J. Organomet. Chem.* **9**, 165 (1967).

38. H. Normant, *Angew. Chem. Int. Ed. Engl.* **6**, 1046 (1967).

39. D. F. Shriver, *The Manipulation of Air-sensitive Compounds*, McGraw-Hill, New York, 1969.

40. G. W. Kramer, A. B. Levy and M. M. Midland, in *Organic Syntheses via Boranes* (H. C. Brown, Ed.), Wiley, New York, 1975, pp. 191–262.

41. *The Use of Aluminium Alkyls in Organic Synthesis; Aluminium Alkyls—Shipping, Handling and Safety; Ethyl's Alkyltainer*. Issued by Ethyl International Co.

42. L. Brandsma, *Preparative Acetylenic Chemistry*, Elsevier, Amsterdam, 1971.

43. H. Schneider, *Z. Anal. Chem.* **135**, 191 (1952).

44. L. H. Horsley, *Azeotropic Data III*, Advances in Chemistry Series No. 116, American Chemical Society, Washington, D.C., 1973.

45. G. A. Fischer and J. A. Kabara, *Anal. Biochem.* **9**, 303 (1964).

46. G. Hesse, *Chromatographisches Praktikum*, Akademische Verlagsgesellschaft, Frankfurt-am-Main, 1968, p. 34.

47. A. J. Gordon and R. A. Ford, *The Chemist's Companion*, Wiley, New York, 1972, p. 375.

48. G. Hesse, *Chromatographisches Praktikum*, Akademische Verlagsgesellschaft, 1968, p. 42.

49. A. H. Gordon and J. E. Eastoe, *Practical Chromatographic Techniques*, Newnes, London, 1964, pp. 51, 82.

50. G. Hesse, *Chromatographisches Praktikum*, Akademische Verlagsgesellschaft, Frankfurt-am-Main, 1968, p. 35; Woelm AG Information Sheet No. 20.

51. J. R. Bower and L. M. Cooke, *Ind. Eng. Chem. Anal. Ed.* **15**, 290 (1943).

52. E. Krell, *Handbook of Laboratory Distillation*, 2nd Edn, Elsevier, Amsterdam, 1963, p. 367.

53. R. L. Augustine, *Catalytic Hydrogenation*, Arnold, London/Dekker, New York, 1965.

54. M. Freifelder, *Practical Catalytic Hydrogenation*, Wiley–Interscience, New York, 1971.
55. P. N. Rylander, *Catalytic Hydrogenation over Platinum Metals*, Academic Press, New York, 1967.
56. F. Zymalkowski, *Katalytische Hydrierungen*, Ferdinand Elke Verlag, Stuttgart, 1965.
57. Ref. 28: e.g. Lindlar's catalyst, Coll. Vol. 5, p. 880; palladium (various forms and substrates), Coll. Vol. 3, pp. 685, 520, 385; Raney nickel W-2, Coll. Vol. 3, p. 181.
58. Ref. 17: e.g. palladium hydroxide/carbon, Vol. 1, p. 782 and Vol. 2, p. 305; special Raney nickel for desulfurization, Vol. 1, p. 729; Urushiba catalysts, Vol. 4, p. 571; iridium tetrachloride/trimethyl phosphite, Vol. 2, p. 229.
59. W. M. Pearlman, *Tetrahedron Lett.* 1663 (1967).
60. *Gmelin's Handbuch der Anorganischen Chemie*, 8th Edn, *Platinum— Part A*, No. 68, Verlag Chemie Weinheim, 1951, pp. 350–362.
61. G. B. Kauffman and L. A. Teffer, *Inorganic Syntheses*, Vol. 7, McGraw-Hill, New York, 1963, p. 233; see also R. Gilchrist, *Chem. Rev.* **32**, 277 (1943).
62. J. W. Mellor, *Comprehensive Treatise on Inorganic and Theoretical Chemistry*, Vol. 15, Longmans Green, London, 1957, p. 595.
63. Manual No. 168, Parr Instrument Co.
64. R. D. Haworth and G. Sheldrick, *J. Chem. Soc.* 1950 (1934).
65. H. Fernholz and H. J. Schmidt, *Angew. Chem. Int. Ed. Engl.* **8**, 521 (1969); K. Sjoeberg, *Acta Chem. Scand.* **22**, 1787 (1968).
66. S. Goldstein and H. J. E. Loewenthal, unpublished, prepared from the corresponding carboxylic acid; C. H. Heathcock and T. R. Kelly, *Tetrahedron*, **24**, 1801 (1968).
67. H. J. E. Loewenthal and S. Shatzmiller, *J. C. S. Perkin I*, 944 (1976).
68. N. Clauson-Kaas and F. Limborg, *Acta Chem. Scand.* **8**, 1579 (1954).
69. Ref. 17, Vol. 1, p. 1022.
70. E. J. Eisenbraun and H. Hall, *Chem. Ind.* 284 (1973).
71. *Chem. Sources* (U.S. and European eds), published annually by Directories Publishing Co., Flemington, N.J., U.S.A.
72. *Chem. Eng. News*, March 17, 1975.
73. IATA Restricted Articles Regulations, issued periodically by International Transport Association, 1155 Mansfield St, Montreal 113, Que., Canada.
74. A. Keren, Frutarom Ltd, Haifa, personal communication.
75. G. W. V. Stark, R. W. White and G. E. Moseley, *Chem. Ind.* 1193 (1971).

Supply Firms Cited

Ace Glass Inc., 1430 Northwest Boulevard, Vineland, N.J. 08360, U.S.A.

The Aldrich Chemical Co. Inc., 940 W. Saint Paul Avenue, Milwaukee, Wis. 53233, U.S.A.

Analtec Inc., Blue Hen Drive, Newark, Delaware 19711, U.S.A. (British associate: Anachem. Ltd, 20a North St, Luton, Beds LU2 7QE, U.K.)

BDH Chemicals Ltd, Poole BH12 4NN, U.K.

Crown-Zellerbach Chemical Products Div., Camas, Wash. 98607, U.S.A.

Drummond Scientific Co.: 'Microcaps' made by this firm are sold by U.S. supply houses such as Arthur Thomas and SGA Scientific; in W. Germany similar product sold by Rudolf Brand, Wertheim/Main-Glashuette (Cat. No. 7091).

E. I. du Pont de Nemours and Co. Inc., Industrial Chemicals Dept, Wilmington, Del. 19898, U.S.A.

Ethyl International Corp., Ethyl Tower, 451 Florida Boulevard, Baton Rouge, Louisiana 70801, U.S.A.

Fluka AG Chemische Fabrik, CH-97470 Buchs SG, Switzerland.

Franz Bergmann KG, 1 Berlin 15, Kurfuerstendamm 170, W. Germany.

GAF Corporation, Calvert City, Kentucky, U.S.A.

Kimble Division, Owens-Illinois, P.O. Box 1035, Toledo, Ohio 43666, U.S.A. (Foreign orders: Overseas Medical Supply Co. Inc., 415 Lexington Ave, New York N.Y. 10017, U.S.A.)

Koch-Light Laboratories Ltd, Colnbrook, Bucks, SL3 OBZ, U.K.

Kontes Glass Co., Vineland, N.J. 08360, U.S.A.

E. Merck, 6100 Darmstadt, W. Germany.

Normalschliff Glasgeraete, 698 Wertheim-Glashuette, W. Germany.

Parr Instrument Co., 211 53rd St, Moline, Ill. 61265, U.S.A. (ask for Manual No. 168).

Pierrefitte-Auby, 46 Rue Jaques Dulud, 92-Neuilly-sur-Seine, France.

Protective Closures Inc., 2207 Elmwood Ave, Buffalo N.Y. 14216, U.S.A.

Sovirel, 27 Rue de la Michodière, Paris 2e, France.

M. Woelm, 244 Eschwege, W. Germany.

Index